Abdunor Zhiyanov
Aider Nazarov

JUSTIFICATION OF BLASTING TECHNOLOGY IN THE QUARRY CONTOUR ZONE

Abdunor Zhiyanov
Aider Nazarov

JUSTIFICATION OF BLASTING TECHNOLOGY IN THE QUARRY CONTOUR ZONE

ScienciaScripts

Imprint

Cover image: www.ingimage.com

This book is a translation from the original published under ISBN 978-620-7-48684-7.

Publisher:
Sciencia Scripts
is a trademark of
Dodo Books Indian Ocean Ltd. and OmniScriptum S.R.L publishing group

120 High Road, East Finchley, London, N2 9ED, United Kingdom
Str. Armeneasca 28/1, office 1, Chisinau MD-2012, Republic of Moldova, Europe
Managing Directors: Ieva Konstantinova, Victoria Ursu
info@omniscriptum.com

Printed at: see last page
ISBN: 978-620-8-52857-7

Contents

The monograph contains theoretical and experimental studies of zoning of instrument massifs of the quarry field, geomechanical substantiation of parameters of quarry sides in zoning of instrument massifs, criterion of zoning of the quarry field on stability of slopes, study of the current state of geomechanical conditions of the rock massifs of the Northern and Central Amantaytau deposits, laboratory studies on determination of engineering-geological characteristics of rocks of the deposit, theoretical substantiation of geotechnical characteristics of the rocks of the deposit, theory of the geotechnical characteristics of the rock massifs of the quarry field.

The monograph will be useful for teachers and students of mining universities and faculties, engineers and technicians of mining enterprises, research and design institutes.

INTRODUCTION

In the world practice in the development of minerals by open-pit mining is characterised by the involvement of fields with complex mining and geological conditions. The analysis of researches of world practice zoning of instrument massifs of a quarry field on stability of slopes and geomechanical substantiation of parameters of sides of quarries at zoning and criterion of zoning of a quarry field on stability of slopes on the given time are studied, not enough. In this connection it is necessary to pay special attention to zoning on division of the quarry field into areas with single conditions of slope stability, drillability, explosiveness, excavability of rocks, determining a unified approach to the design and organisation of mining operations.

To date, the world is conducting scientific research on the zoning of the karer field on the explosiveness of rocks and can be carried out on the basis of geological and structural mapping with the involvement of materials of detailed exploration of deposits. In this regard, special attention is paid to the strength of rocks is determined by their belonging to one or another engineering-geological lithotype, and within one lithotype it is taken into account in the gradation of rocks by their blockiness. The most important issue in zoning is the choice of zoning criteria - one or several values to divide the object into areas within which the conditions are relatively the same. Thus, the stability coefficient equal to unity should be taken as a boundary value for dividing the slope of the carer into stable and unstable areas, at which the holding and shear forces on the slope are equal, less than unity - the slope is not stable. Development of methods of mining geometric zoning of karer fields, theoretical justification of physical and mechanical parameters of the rock mass, the establishment of limit parameters of the angle of slope of the sides and the technology of blasting in the instrumental zone of karer on the technical conditions of development to improve the efficiency and safety of mining operations is an urgent problem of science and practice of mining production, the solution of which contributes to improving the economic efficiency of enterprises.

Significant contribution to the development of the fundamentals, the development of the theory and methodological issues of variability assessment and modelling of various indicators of deposits, assessment of geomechanical conditions and issues of assessment of mining-technological conditions of development - drillability, explosivity, excavability of rocks and zoning of quarry sides by stability were made by scientists P. P. Bastan, V.A. Bukrinsky, G.I. Vilesov, V.M. Gudkov, A.B. Kalichenko, A.B. Kamdan.P. Bastan, V.A. Bukrinsky, G.I. Vilesov, V.M. Gudkov, V.M.

Kalichenko, A.B. Kamdan, A.U. Margolin, V.F. Myagkov, M.V. Ratz, P.A. Ryzhov, P.K. Sobolevsky, I. N. Ushakov, V.N. Ushakov, V.N. Ushakov, V.N. Ushakov, and P.K. Sobolevsky.N. Ushakov, V.S. Khokhryakov, L.I. Chetverikov, B.G. Afanasyev, A.M. Galperin, A.M. Demin, V.G. Zoteev, A.I. Ilyin, V.A. Mironenko, R.P. Skatov, M.E. Pevzner, V.N. Popov, I.I. Popov, S.I. Popov, V.D. Polovovov, V.T. Sapozhnikov, Yu.I. Baron, Y.I. Belyakov, O.N. Golubintsev, B.N. Kutuzov, V.I. Molchanov, V.I. Mosinets, B.R. Rakishev, V.K. Rubtsov, V.V. Rzhevskii, I.A. Tangaev, etc.

In the Republic of Uzbekistan a significant contribution was made by V.R. Rakhimov, B.R. Raimzhanov, S.S. Saidkasimov, K.D. Salamova, F.Y. Umarov, S.S. Zairov, Z.S. Nazarov and others, they have achieved significant success in the study of the geomechanical state of the rock mass during the development of mineral deposits by open pit mining.

However, up to the present time, the lack of links between the indicators of engineering-geological zoning with the parameters of geomechanical, technological processes of mining production and methods of mining-geometric zoning of quarry sides by stability and by physical and technical conditions has not been studied, which limits the use of geometricisation results in the design and operation of quarries.

Thus, the zoning of sides on stability taking into account geomechanical conditions of the rock mass and justification of blasting technology at the contour zone of the open pit is an important direction to ensure the efficiency and safety of mining operations.

CHAPTER 1

ANALYSIS OF STUDIES OF WORLD PRACTICE ZONING OF INSTRUMENT MASSIFS OF QUARRY SLOPE STABILITY FIELDS

§ 1.1. Analysis of the world practice of zoning of instrumental quarry field massifs

Rezoning of physical and technical conditions of development is a set of operations performed to assess and forecast the stability of slopes, drillability, explosiveness, excavability of rocks in different parts of the open pit field. The main operations (elements) of zoning are: selection of zoning indicator (index); determination of informative parameters (influencing factors); establishment of relationship between informative parameters and zoning index; calculation of zoning indicator; selection of zoning criterion and preparation of forecast map.

Physical and technical development conditions are determined by interaction of two groups of factors - natural and technological. The most important of the natural factors affecting the stability of open pit slopes and technological processes of open pit mining are physical and mechanical properties of rocks, structural and tectonic structure of the massif, hydrogeological and seismic conditions.

In order to assess the geomechanical and technological conditions of development, it is necessary to conduct research to establish the nature of spatial variability of natural factors, mathematical modelling and mapping of their location in the massif.

The works of P.P. Bastan, V.A. Bukrinskiy, G.I. Vilesov, V.M. Gudkov, V.M. Kalichenko, A.B. Kamdan, A.M. Margolin, V.F. Mykolaev, and V.F. Kamdan are devoted to the development of theoretical and methodological issues of assessing variability and modelling the location of various field indicators. Bukrinsky, G.I. Vilesov, V.M. Gudkov, V.M. Kalichenko, A.B. Kamdan, A.M. Margolin, V.F. Myagkov, M.V. Rats, P.A. Ryzhov, P.K. Sobolevsky, I.N. Ushakov, V.S. Khokhryakov, L.I. Chetverikov, etc. [1-12].

The peculiarities of the initial information on physical and mechanical properties of rocks obtained at the stage of field exploration (limited volume, irregularity of the location of sampling points, multiplicity of values of the property indicator in each point of sampling), make it difficult to use a number of traditional methods of studying spatial variability. Therefore, it is necessary to develop a special methodology for mining geometric analysis of the location of these property indicators.

Analysis of studies on the issues of assessment of geomechanical conditions of development conducted by B.G. Afanasiev, A.M. Galperin, A.M. Demin, V.G. Zoteev, A.I. Ilyin, V.A. Mironenko, R.P. Skatov, M.E. Pevzner, V.N. Popov, I.I. Popov, S.I. Popov, V.D. Polovovov, V.G. Sapozhnikov. Pevzner, V.N. Popov, I.I. Popov, S.I. Popov, V.D. Polovovov, V.G. Sapozhnikov, Yu. [13-23] showed that at present the most well developed methods for assessing the stability of quarry slopes are based on the theory of limit equilibrium. For the purposes of zoning of quarry fields according to geomechanical conditions of development, the point system of slope stability assessment represents a wide range of possibilities. Its essence consists in the fact that each factor influencing the slope stability receives a certain score, and the sum of scores from all factors characterises the slope stability in specific conditions of field development. The disadvantages of the known scoring methods of stability assessment are: incomplete consideration of the main factors affecting stability; insufficient validity of the numerical values of scores for different factors; impossibility to determine slope stability indicators (stability reserve factor) or parameters of stable slope (slope angle, height).

In this regard, the development of a reliable scoring system for assessing slope stability based on the theory of limit equilibrium and a methodology for zoning geomechanical conditions remain important research challenges.

Studies on the issues of assessment of mining-technological conditions of development - drillability, explosiveness, excavability of rocks were conducted by L.I. Baron, Y.I. Belyakov, O.N. Golubintsev, B.N. Kutuzov, V.I. Molchanov, V.N. Mosinets, B.R. Rakishev, V.K. Rubtsov, V.V. Rzhevsky, I.A. Tangayev and others. Rzhevskii, I.A. Tangaev, and others. [24-32], based on the results of the analysis of their works, it was established that as indicators of zoning it is advisable to use such characteristics of technological properties, the determination of which is possible on the basis of dependencies on physical and mechanical properties of rocks. This will make it possible to carry out a predictive assessment of development conditions based on the results of geological exploration.

The authors of the paper Rybin V.V., Potapov D.A., Kalyuzhny A.S. [33] carried out studies on zoning of the open pit field of the Oleniy Ruchey deposit by depth using the geomechanical classification of Professor D. Lobshir.

The Oleniy Ruchey deposit is located in the south-eastern part of the Khibiny massif of the Russian Federation. The ore bodies of the deposit have a multilayered structure, and the overlying side of the deposit is in contact with ricechorrites, in the zone of conjugation with khibinites. In addition to ricechorrites, foyaites occur on the hanging wall side of the deposit. In addition to these main types of host rocks, lenses of apatite-nepheline ores alternate with gneiss-like urtites, ijolites, melteigites, and other host rocks [34].

The largest structural fault in the Oleniy Ruchey deposit is the Main Fault. The main fault is north-east trending, north-west dipping with angles of approximately 40 to 45^0. The Main Fault is the main factor controlling the variability of parameters of structural heterogeneities of the rock massif within the quarry field [34].

At the stage of detailed exploration and refinement of reserves in the mid-1980s of the last century, the physical and mechanical properties of ores and host rocks were determined by examining samples made from cores of deep (more than 1 km) wells [35]. Indicators of some physical properties of rocks are given in [36].

These data, in particular the high values of tensile strengths, indicate that all the main types of host rocks belong to the category of strong rocky rocks. The brittleness coefficient (aszh/ar ratio) in all cases is greater than 10, which indicates the tendency of these rocks to brittle fracture. The value of wave velocity in the rocks ranges from 2.3 to 5.8 km/s, which is also characteristic of strong rocky rocks.

In 2010, additional studies of the physical properties of the main rock types in the upper part of the deposit were carried out to assess the stability of the planned open pit at the Oleniy Ruchey deposit [37].

From analysing the data on the physical properties of rocks, it is obvious that the values of strength indices determined in the 1980s are significantly higher compared to the 2010 data. The most probable reason for this is that in the 2010 determinations sampling was carried out in the upper part of the deposit, i.e. in weathered and fractured rock formations.

It is also found that with depth the disturbance and fracturing of the rock massif decreases, which leads to higher parameters of physical and mechanical properties of rocks in the massif.

To take into account the influence of stress-strain state parameters on the stability of the instrumental massif, the natural stress state of the field rock massif was predicted. As a result, three characteristic zones were identified by the level of acting stresses [36]:

I - weakly stressed zone ($\sigma_3 < 20$ MPa), from the surface to a depth of 400 m;

II - medium stressed zone (20 MPa $< \sigma_3 <$ 40 MPa), from 400 to 1000 m;

III - highly stressed zone ($\sigma_3 > 40$ MPa), over 1000 m.

According to the indicators of core fracturing and the character of its disking,

3 zones were also distinguished according to the depth of the deposit [35]: up to 400 m; from 400 to 1000 m, more than1000 m.

The upper zone up to 400 m depth is characterised by increased fracturing, which is most likely due to the presence of a weathering zone, i.e., a zone of weakened rocks noted on the basis of seismic surveys. The values of acting stresses in this zone are significantly lower than at greater depths.

In the second zone, at a depth of 400 to 1000 m, core disking is manifested, which is confined mainly to zones of high-strength rocks (urtites, ijolites). The rocks in this zone are less fractured.

Within the third zone, at depths greater than 1000 m, core disking is evident at much greater intervals, indicating a higher value of active stresses in the rock massif and a much lower intensity of fracturing.

On the basis of available materials on the geomechanical condition of the deposit it is possible to determine the geomechanical rating of the rock massif. Recently, the geomechanical rating classification MRMR (hereinafter RL) developed abroad by Prof. D. Lobshire, which has been widely used in Western countries since the mid-70s of the last century, has become very popular [38, 39]. The methodology makes it possible to determine at a preliminary stage the main parameters of mining operations based on the results of geological exploration, which is very useful for preliminary assessment of the stability of the sides of open pits.

The algorithm for determining the RL rating indicator is shown in Fig. 1.1. The scheme shows that the value of the RL rating is determined by the sum of partial ratings that take into account the strength characteristics of the massif, quantitative and qualitative characteristics of fracturing, which are further multiplied by correction factors of weathering, crack orientation, stress state, hydrogeology, etc. The *RL* rating value is determined by the sum of partial ratings that take into account the strength characteristics of the massif, quantitative and qualitative characteristics of fracturing, which are further multiplied by correction factors of weathering, fracture orientation, stress state, hydrogeology, etc.

The value of the RL rating is expressed by the following formula:

$$R_L = \left(R \cdot \sigma_{бл} + R_{кт} + R_{ут}\right) \cdot k, \qquad (1.1)$$

where $R\sigma bl$ - strength of the rock block; RKT - rating by number of cracks; R_{ym} - rating of fracture conditions; k - coefficients taking into account weathering, orientation of cracks, stresses in the massif, blasting, presence of underground water inflows.

Figure 1.1. Algorithm for determining the *RL* rating indicator according to the classification of Professor D. Lobshire

In accordance with Table 1.1 developed by Professor D. Lobshire, the results of the *RL* rating calculation show that the rocks of the Oleniy R*uchey* deposit in the upper part of the deposit up to a depth of 100 m, except for the overlying moraine deposits, belong to Class 4 and are characterised by low stability. The rocks at a depth of 100 to 300 m, when subjected to low tectonic stresses, belong to the third class and are characterised by medium stability, but when subjected to high tectonic stresses they will belong to the second class and are characterised by good stability. At a depth of 300 and more metres the rocks have good stability and belong more to the second class.

Table 1.1.

RL rating values according to rock class by Lobshear

RL Rating	81-100	61-80	41-60	21-40	5-20
Lobshire breed class	1	2	3	4	5

A. Hines and P. Terbrugge [40] developed recommendations for selecting approximate values of pit slope angles based on the calculated MRMR rating. These recommendations are presented in Table 1.2.

Table 1.2.

Approximate values of pit wall slope angles according to rock class according to rock class

Breed class	1	2	3	4	5
Slope angle of the	75°±5°	65°±5°	55°±5°	45°±5°	35°±5°

quarry face					

Based on the calculated *RL* rating of the Oleniy Ruchey deposit rocks for the open pit, the following options of side slope angles can be recommended for consideration as a first approximation:

- for depths up to 100 metres - 35-500 ,
- for depths from 100 to 300 metres - 50-600,
- deeper than 300 metres - to 65^0.

Graphically, the zoning of the rating results by depth is shown in one of the transects, Figure 1.2.

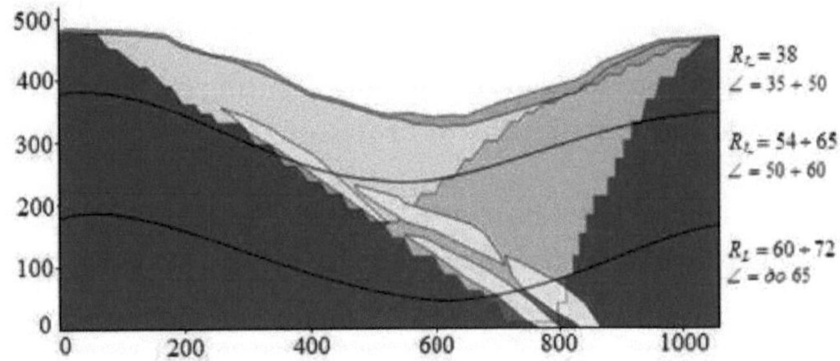

(orange - layer of moraine sediments, blue and red - possible pit outlines, light green - orebodies, black - ore bodies).
possible pit contours, light green - ore bodies, black - black - conditional boundary of the rating indicator division, dark grey - host rocks). host rocks)

Figure 1.2. Section 32+50 m with the position of conditional boundaries of ratings

It should be noted that the calculation of *RL* ratings did not take into account the type of drilling and blasting operations. The use of contour blasting will contribute to the correspondence of the real state of the near-contour part of the rock massif to the given results of the calculation of ratings.

Assessment of the stability of the instrument massifs of deep open pits and their zoning by the stability factor are considered in the works by Nizametdinov F.K., Urdubaeva R.A., Ananin A.I. and Ozhigina S.B. on the example of Sokolovskiy open pit of JSC "SSGPO" [41]. [41].

To assess the stability of quarry slopes in VNIItsvetmet, a software package "BORT" has been developed. The algorithm of the software complex "BORT" is based on the theory of limiting equilibrium of rocks [42-44], according to which the violation of stability of the quarry slope occurs in the form of collapse or sliding of rocks on the sliding surface, which is a

combination of rectilinear and curvilinear sections.
In accordance with the accepted provisions, a slip surface is constructed for a given point on the side of the pit or scarp and the stability factor n is calculated for it, which is calculated as the ratio of the sums of shear and restraining forces acting on the slip surface. In the specified range of the slope surface under consideration, the dependence $n=f(Bi)$ is plotted and the minimum stability factor and the width of the possible collapse prism are determined.
The BORT software package implements the method of direct enumeration of all possible positions of the slip line with a fixed incremental step. The calculation scheme developed at KSTU [45] by Prof. Popov I.I., Prof. Shpakov P.S., Prof. Poklad G.G. and others was used as a basis for the development of the computer simulation programme.
The software package is developed in the Delphi programming environment designed for use on personal computers in Windows-95 and higher.
The software system allows to assess the stability of any configuration of the board and calculate the slope stability factor at any point of the board at a given contour and certain mechanical characteristics of rocks (cohesion, angle of internal friction, volumetric weight).
The peculiarity of the software complex "BORT" is that its algorithm allows to solve not only a flat problem within one section of the quarry face, but also to obtain a volumetric solution when the calculation results are used together with the geoinformation system Surpac.
Development of "BORT" software system capabilities is the development of a number of technologies for its joint use with Surpac geoinformation system both for preparation of high-quality initial data for calculations and for group processing of calculation results for individual sections. In particular, one of the developed technologies makes it possible to present a series of calculation results for individual sections in the form of a map of pit stability, which makes it possible to perform zoning of pit sides by stability factor.
The main input data for the assessment of the stability of quarry sides using the software complex "BORT" are weighted average indices of volumetric weight, cohesion and angle of internal friction included in the calculated cross-section of the rocks of the quarry apparatus massif.
When assessing the stability of the appurtenant massifs of the Sokolovskoe open pit, a fan of sections was used, uniformly covering the quarry face and, if possible, located perpendicular to the ledges of the face with the most typical design sections of the quarry face Fig.1.3.

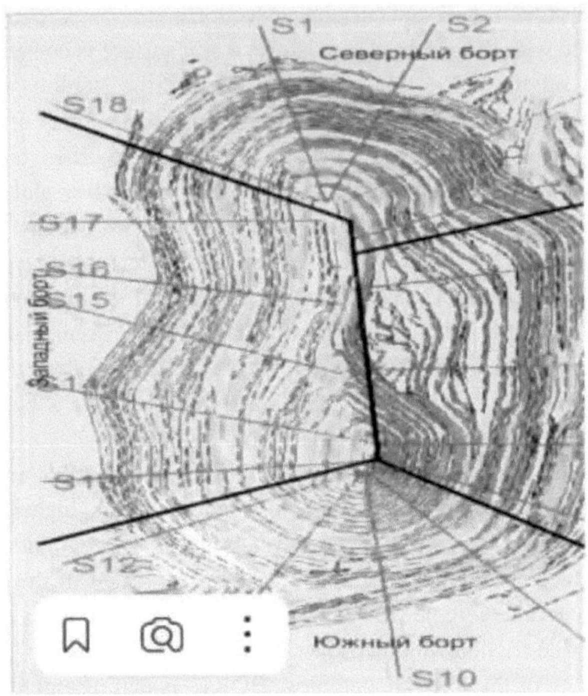

Fig.1.3. Calculated sections of the Sokolovskoye quarry face

When exporting the calculation results of the BORT software system for all calculated cross sections to the Surpac geoinformation system, a combined model of the Sokolovskoye open pit was obtained, shown in Figure 1.4.

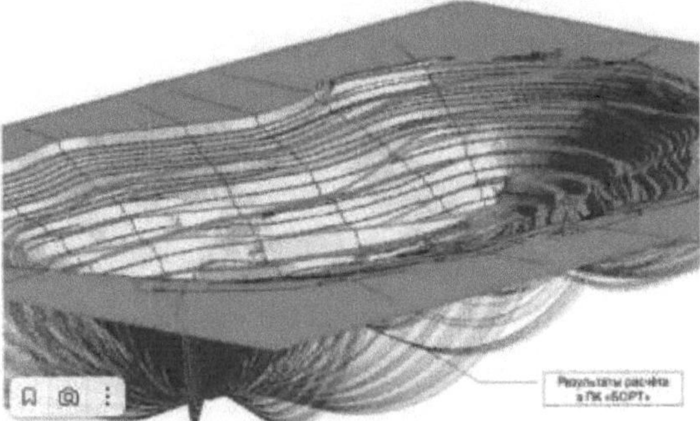

Fig.1.4. Combining the volumetric model of the quarry with the results of calculations in the "BORT" software package

The obtained digital merged model was used in development of the pit stability map. This map allows to assess the decrease in the stability of the quarry sides, identify the weakest areas, develop and model measures to improve their stability. At the same time, it is possible to estimate the scope of works to create a stopper prism or to unload the sides by flattening the quarry slopes, etc.

The safety factor for all calculated sections exceeds 1: the minimum value of the safety factor is 1.11 for section *S11* on the south side of the quarry.

The authors Dunayev V.A., Serem S.S., Gerasimov A.V. and Absatarov S.H. developed a methodology for zoning of quarry fields by blockiness and explosivity of rocks [46].

The authors' long-term studies at the operating deposits of ferruginous quartzites in the KMA basin have shown that their zoning by rock explosiveness can be performed on the basis of geological and structural mapping of operating quarries with the involvement of detailed exploration materials of the deposits. In this case, the strength of rocks is determined by their belonging to one or another engineering and geological lithotype, and within one lithotype it is taken into account in the gradation of rocks according to their blockiness.

In complex folded massifs of metamorphic rocks, blockiness is created mainly by fractures of three mutually orthogonal systems *M* - along the layering of rocks, *N* - transverse to their strike, *K* - subparallel to the strike of rocks and subperpendicular to their layering.

In view of the above, it is necessary and sufficient to measure the orientation of cracks of different systems and the distance between cracks in each system during in-situ fracturing studies, which avoids errors in the interpretation of the results obtained.

The determinants of explosive fracture resistance of metamorphic rocks (blockiness and strength) are caused by the same factors and change in a coordinated manner under their influence, revealing a direct correlation between them.

Firstly, the lithological factor is manifested in the regular increase of blockiness and strength of metamorphic rocks in the following sequence: shales - quartzite sandstones - ferruginous quartzites ore - ferruginous quartzites low-ore.

Secondly, all rock types are characterised by an increase in their blockiness and strength in the direction from the wings to the fold locks. This phenomenon reflects the influence of the structural factor. An explanation of its origin is given in [47].

Third, the hypergenic factor caused an increase in blockiness and rock strength with depth. Physical weathering and hypergenic unloading of the massif led to the above-mentioned directional change in blockiness, and chemical weathering led to the change in rock strength.

Direct connection of blockiness and strength of metamorphic rocks is characteristic for deposits of ferruginous quartzites of the Krivoy Rog basin.

Two important conclusions follow from the established relationship between blockiness and rock strength in the massifs of ferruginous quartzite deposits:

- provided that the array is well studied in terms of rock fracturing (blockiness), their strength (within the engineering-geological lithotype) is automatically taken into account in the classification and scheme of the array zoning by rock strength, i.e. these classification and scheme adequately reflect the gradation of rock explosivity categories and their placement in the quarry field. Thus, it is possible to avoid a large volume of expensive physical and mechanical tests of rock samples, which would be necessary to perform for reliable zoning of the massif by rock strength, given its high variability within the lithotype.
- quarries developing deposits of ferruginous quartzites are favourable objects for the introduction of an automated system for determining the explosiveness of rocks in the process of drilling by specific energy consumption of roller cone drilling, since this indicator is closely correlated with the strength of rocks [48], which in deposits of ferruginous quartzites is directly related to their blockiness.

Application of such system will allow to specify the scheme of zoning of the quarry field on explosiveness of rocks built on the results of field researches of massif blockiness as the data on specific energy consumption of roller drilling are accumulated and processed.

The general methodological scheme of zoning the quarry field by rock explosivity is as follows Fig. 1.5.

The implementation of the proposed methodology was carried out on the example of Lebedinsky KMA deposit.

The criterion values of the average size of separateness for different categories of rock blockiness are taken in accordance with the "Temporal classification..." [49], in which for the purpose of more differentiated geometrization of the massif by the degree of rock fracturing, the most widespread category III in the field was divided into two subcategories: III-a (0.5-0.75) and III-b (0.75-1.0 m).

The classification of rocks by explosivity is based on two criteria: belonging to a certain engineering-geological lithotype and blockiness category Table 1.3.

Fig. 1.5. Scheme of zoning of the quarry field by blockiness and explosiveness of rocks

The scheme of zoning of the quarry field by rock explosiveness the explosiveness map was built on the basis of a similar scheme by rock blockiness, taking into account the distribution of different engineering-geological lithotypes.

To keep the blast map up to date, the geological and structural plan of the open pit should be updated as mining progresses and the boundaries of the different blast categories should be adjusted based on this plan. Such adjustments will be more reliable if blasting conditions and quality are systematically analysed, and additional in-situ geological observations and measurements of rock blockage are made based on the results.

Classification of rocks on explosivity and the map of explosivity of the quarry field are the main documents of the standard project of drilling and blasting works at the quarry. The availability of these documents allows to automate the design process of drilling and blasting works [50].

Table 1.3 **Classification by Explosivity of Rocks of Lebedinsky deposit**

Explosion category	**Engineering and geological lithotypes**	**Fracture category**	**Well grid*, *m***		**Specific BB** consumption rate, *kg/m³***
			a	***b***	
1 - Exceptionally explosive	Weathered shales and quartzite sandstones Oxidised ferruginous quartzites	I -II I	9	8	0,4 - 0,5

2 - highly explosive	Oxidised and semi-oxidised ferrous Quartzites Unweathered quartzitic sandstones and shales	II II - III -a	8 - 7	7	0,5 - 0,6
3 - lightly explosive	Unoxidised ferruginous quartzites Oxidised and semi-oxidised ferruginous quartzites ferruginous Quartzites Unweathered quartzitic sandstones and shales	II III III-b to IV	7	6 - 7	0,6 - 0,8
4 - medium explosive	Unoxidised ferruginous quartzites	III-a	6	5 - 6	0,8 - 1,0
5 - difficult-to-explode		III-b	5,5	5,5	1,0 - 1,2
6 - very difficult to detonate		IV - V	5,0	5,0	1,2 - 1,4
Note: • - distance: *a* - between rows of wells, *b* - between wells in rows. •• - for standard BBs.					

Authors Koltsov P. V., Ivanov Y. S., Palutina E. N. and Andreev O. H. [51] have evaluated the stability of the Uchalinsky open-pit mine sides. At the same time, ensuring the stability of open-pit mine sides and temporary prevention of emerging slope deformations under changing mining and geological conditions are the most important tasks at mining enterprises.

The purpose of the conducted research was to zonation of the quarry sides by the degree of stability and assessment of the possibility of safe production of processes.

Conducting a stability assessment of the Uchalinskoe quarry faces was one of the main tasks.

The stability of the sides of the Uchalinsky quarry was assessed using ten design profiles according to the methodological guidelines of VNIMI [51].

Based on the results of the calculations, the pit sides were zoned according to their stability. The stability reserve coefficient was taken as a zoning index.

The degree of stability of the quarry sides based on the normative documentation [51] was taken as a criterion for zoning. At the same time, the following categories were identified:

1. The value of the reserve coefficient n> 1.3; the instrument massif experiences mainly elastic deformations, the values of which are within the accuracy of surveyor measurements; relative horizontal deformations do not exceed $1\text{-}10^{-3}$; the rim is in stable condition.
2. The value of the stock factor is 1.2 < n < 1.3; cracks appear; the total

displacements of the surface of the instrument massifs reach 200:300 mm; the relative horizontal deformations can reach (2:5)-10^{-3}; the horizontal component of the shear vector is predominant; the displacements are damped in time.

3. The magnitude of the stock factor is 1.1 < n < 1.2; zakols appear; horizontal deformations can reach 30-10^{-3}; and total displacement values are 1.5:2 m; deformations are predominantly damped in time.

4. The value of the safety factor is 1.05 < n < 1.1; further development of dangerous deformations occurs.

5. The value of the reserve coefficient n < 1.05; in the short term, the board slides or collapses; however, in case of significant movements of the instrument surface and slope deformations, the board may retain short-term stability under the condition of changes in the mining and mining-geological condition of the board.

§ 1.2 Geomechanical justification of the parameters of quarry sides at zoning of instrumental massifs

At geomechanical substantiation the methods and ways of estimation of stability of sides and ledges of quarries at all stages of development of deposits are given [52].

Calculation methods are of recommendatory nature. The composition and types of works performed on geomechanical substantiation of the parameters of the sides and ledges of quarries are considered in accordance with the requirements of the rules in the field of industrial safety [53].

The structure and stages of works to ensure the stability of quarry sides and ledges are presented in the figure below Fig. 1.6.

Principles of selecting an analogue deposit at the early stages of development of deposits in the absence of reliable data on the rock massif, it is allowed to determine the parameters of the sides and ledges of open pits by the method of analogies.

The method of analogies can be used to determine physical and mechanical properties of the rock massif, parameters of the sides and ledges of open pits.

The basic requirements of an analogy:

а) Analogy can be carried out either by one or by a number of attributes, the more attributes, the more reliable the analogy;

б) the analogy must be made on the merits and not on a formal basis;

в) the same attributes must be compared;

г) it is necessary to separate attributes by importance by weight. Differences in the most significant features, but convergence in features with low weights may indicate that the analogy is incorrect;

e) the attributes should be the most diverse and versatile;

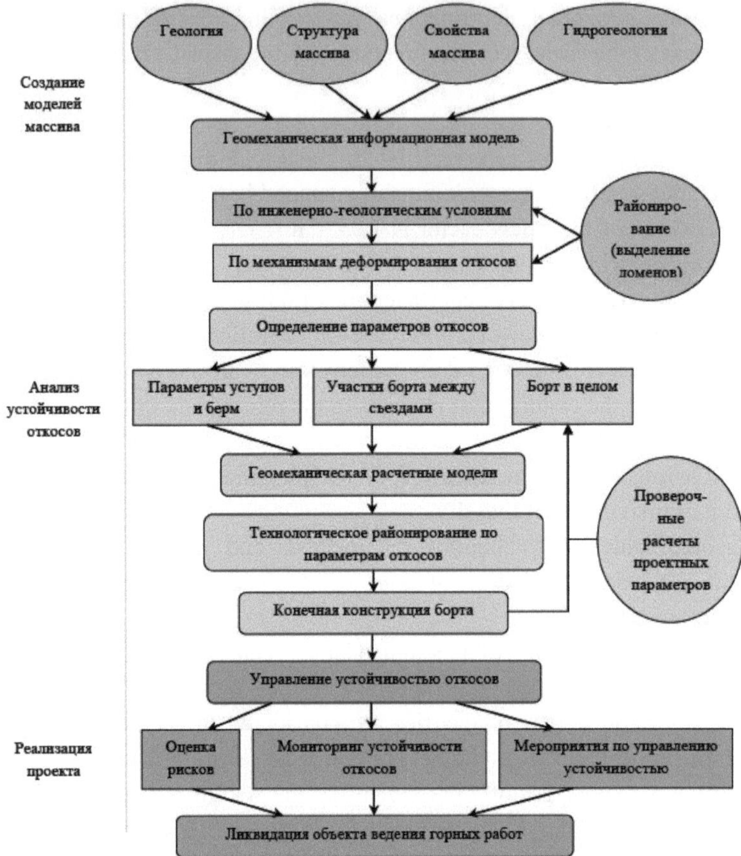

Fig.1.6. Structure of geomechanical support of stability of sides and **ledges**

(e) When transferring a parameter from a studied field to a target field, its overall relationship to a number of key attributes should be assessed.

The following attributes may be used for analogy:

а) regional situation;

б) conditions of deposit formation;

в) type of mineral;

г) rock description, rock type and strength, inclusions and impurities, contrast, rock types in the rock massif;

д) structure of the deposit, presence of structural floors, forms of monoclinal, folded, blocky and hollow, inclined, steep rock strata;

е) intensity of the processes of rock decompaction and weathering;

ж) hydrogeological conditions: conditions of groundwater formation, number and thickness of aquifers, boundary conditions;

3) climatic conditions.

Failure criteria are used to describe the limit state of rocks when calculating the stability of the sides and ledges of quarries. The main fracture criterion is the linear Coulomb-Mohr criterion:

$$\tau = \sigma \cdot tg\varphi + C$$

$$\tau = \sigma \cdot tg\varphi' + C' \qquad (1.2)$$

where τ - tangential stress, MPa; σ - normal stress, MPa; φ, C - angle of internal friction and cohesion in the rock; φ′ - angle of friction on the weakening surface, deg; C' - cohesion on the weakening surface, MPa.

The Coulomb-Mohr fracture criterion assumes the use of a linear approximation of the ultimate envelope of the strength passport

Figure 1.7.

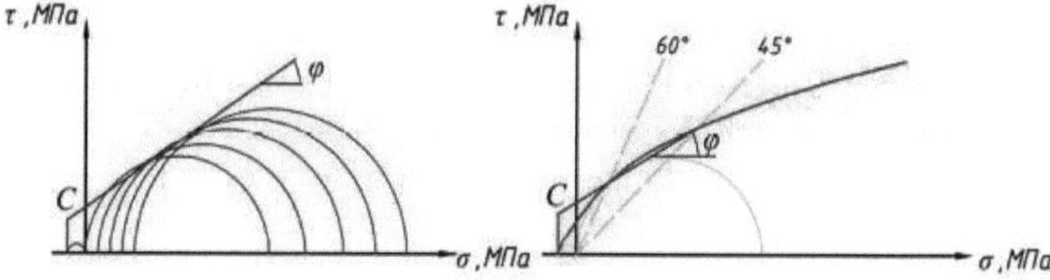

Fig. 1.7. Linear approximation of the results of laboratory tests of rocks (Coulomb-Mohr criterion)

Non-linear array strength criteria are applied at the discretion of a specialised organisation.
To assess the stability of slopes in rock massifs, the nonlinear Hooke-Brown fracture criterion is used Fig. 1.8, [54-56]. Application of this fracture criterion for dispersed rocks is not acceptable.
The generalised Hooke-Brown failure criterion:

$$\sigma_1' = \sigma_3' + \sigma_{ci}\left(m_b \frac{\sigma_3'}{\sigma_{ci}} + s\right)^a \qquad (1.3)$$

where *aci,* - uniaxial compression strength in the sample in the Hooke-Brown criterion; *t is* the constant of the rock massif, which is determined by the formula:

$$m_b = m_i \cdot \exp\left(\frac{GSI-100}{28-14D}\right) \qquad (1.4)$$

mi - undisturbed rock constant; *s* and *a* - rock constants calculated by the following expressions: where *GSI* (geological strength index) - geological strength index; *D* (disturbance factor) - blasting disturbance factor.

$$s = \exp\left(\frac{GSI-100}{9-3D}\right) \qquad a = \frac{1}{2} + \frac{1}{6}\left(e^{\frac{-GSI}{15}} - e^{\frac{-20}{3}}\right) \qquad (1.5)$$

Factor *D* varies from *0* - undisturbed by blasting, to *1* - very highly disturbed by blasting.

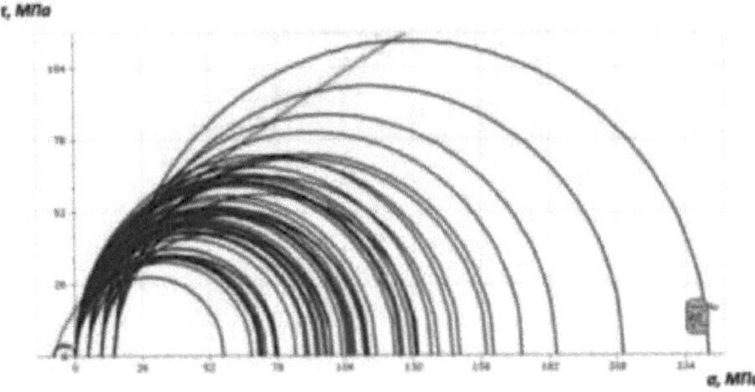

a) strength passport in coordinates of normal and tangential stresses (σ, τ),

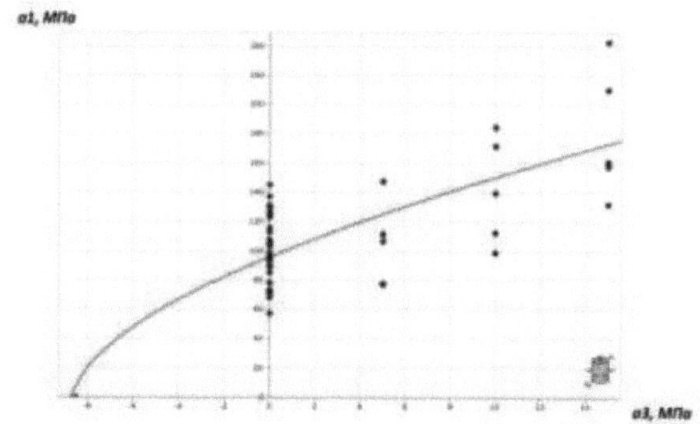

б) strength passport in coordinates of principal stresses ($\sigma 1$, $\sigma 3$).

Figure 1.8. Nonlinear approximation of the results of laboratory Rock tests (Hooke-Brown criterion)

The values given in Table 1.4 [54-56] are used to select the factor *D*.

The GSI geological strength index is used only for rocky and semi-rocky fractured rocks and takes into account the structural structure of the massif and the characteristics of the contacts. Also, the application of *GSI* is limited by the scale effect of Fig. 1.9, where a high level of anisotropy can be observed at a certain scale level. In such cases, the Coulomb-Mohr properties (cohesion and contact friction angle) are specified along the anisotropy direction.

Table 1.4.

Determination of *D* disturbance factor by blasting operations

Appearance of the array	Description of the impact on the array	Proposed value of *D*
	Undisturbed rocks - laboratory strength Use Blasting using gentle blasting techniques that minimise disturbance to the surface massif	***D*** = *0,7*
	One fracture system - high anisotropy - DO NOT use GSI Explosion without the use of gentle technologies	***D***= 1.0

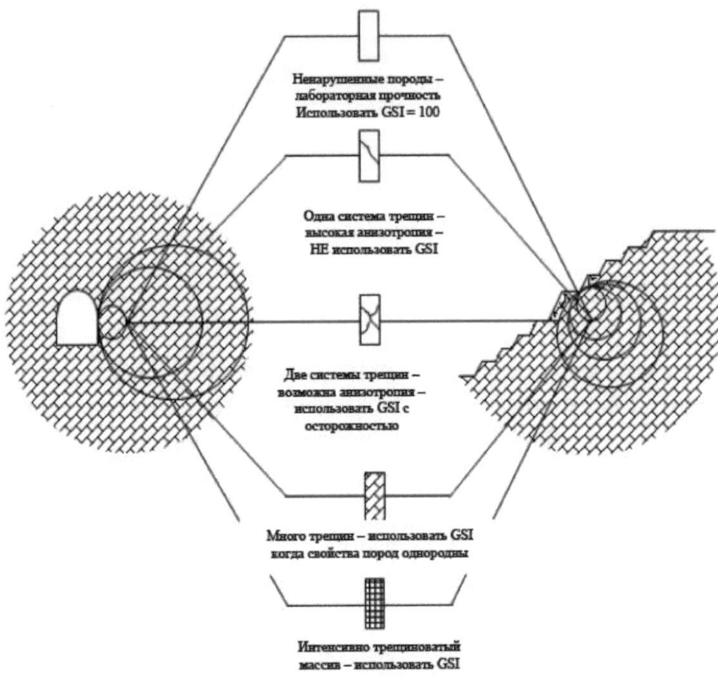

Two fracture systems - anisotropy possible - use GSI with caution
Many fractures - use GSI when rock properties are homogeneous

Fig.1.9. Boundaries of *GSI* parameter application

Regionalisation by possible mechanisms of deformation of the instrumental massif (kinematic analysis) is carried out on the basis of domains identified by natural factors, taking into account the location of the quarry, dump. One and the same area, allocated by natural factors, for slopes with different strike can deform according to different schemes. In this regard, the natural area is divided into design areas, in which different mechanisms of slope deformation are assumed. The zoning by possible mechanisms of slope deformation can be performed both in graphical and textual form.

The selection of computational areas is scale-aware and should be divided into the following hierarchical levels:

- scale of the scarp (weakening surfaces are represented by fine-order structures - fracture systems, layering);
- scale of the rim section (weakening surfaces are represented by structures of higher order - extended fracture systems, layering, faults);
- scale of the rim (weakening surfaces are represented by high-order structures - large faults. The strength of the massif, the direction of layering,

and the water content of the rim are also of fundamental importance).
Zoning [57] is permitted for the design of the outline of quarries:

- by pit wall design (by parameters of ledges and berms);
- on the parameters of the pit sides (including ledge group).

To carry out zoning by structural parameters of quarries, it is necessary to use the areas allocated by mechanisms of slope failure at an appropriate scale.
The boundaries of zoning by natural factors and by mechanisms of destruction of the instrument massif do not always coincide with the boundaries of zoning by structural parameters of the open pit. In the same domain, different slope design parameters can be obtained at different azimuths of ledges and sides.
After the calculations for all these areas have been carried out, the selection and delineation of slope parameters for the entire quarry field is carried out.
The topography and depth of field development may influence the delineation of areas based on side parameters.
In strong rock formations for zoning according to the parameters of the quarry sides, it is allowed to use the structural general angle of the quarry side, calculated according to the parameters of ledges and berms.
Constructive general angle of the quarry face is determined by the formula:

$$a_z = arctg\frac{H}{\sum B_i + \sum h_i ctg\alpha} \quad (1.6)$$

where ΣBi - sum of lengths of all horizontal berms (safety and transport berms), *m*, $\Sigma hictga$ - sum of lengths of all scarp slope deposits in horizontal projection, *m*.
This formula is used to calculate technically acceptable angles when they are smaller than the general side slope angles justified by stability calculations.
The main results of zoning by constructive parameters of quarries are the allocation of quarry field sectors with the same parameters of slopes.

§ 1.3 Criterion for zoning the quarry field by slope stability

When assessing the stability of quarry slopes, three types of problems are usually solved [58]:

1. Determination of the maximum allowable slope height at a given slope angle and normative stability coefficient.
2. Determination of the slope angle of a stable slope at a given height and normative stability coefficient.
3. Determination of stability coefficient at given slope angles and slope heights.

The first two types of tasks are mainly solved at the stage of pit design or during the design of subsequent stages of its construction and are aimed at establishing geometric parameters of stable sides and ledges for efficient and safe mining of the deposit. The third type of tasks is solved at the stage of field operation in order to identify potentially dangerous sections of the sides for the organisation of instrumental observations of stability and development of anti-strain measures or to identify sections of the side where the slope inclination angle can be increased to reduce the volume of stripping works.

All three types of problems can be solved on the basis of the score system of stability assessment [59], which is also used in the predictive zoning of quarry fields by the factor of slope stability.

Both slope height or slope angle, and stability coefficient can be taken as a zoning indicator (an integral value, which collectively reflects the property of interest of the object being zoned, and the value of which is used to select areas of the object with relatively typical conditions). The adoption of one or another parameter as a zoning indicator does not introduce fundamental changes in the zoning and depends on the research objective.

For example, let's focus on the zoning of the quarry field by the factor of stability in the process of quarry operation.

In this case, it is reasonable to take the stability reserve coefficient as an indicator of zoning. From natural factors influencing stability will be: density, adhesion and angle of internal friction of rocks composing different sections of the quarry sides. Data on tectonic disturbances dip angle of weakening surface, adhesion and angle of internal friction on it; the degree of watered slope, seismicity coefficient of the field development area, from technological factors height and angle of inclination of the side.

As initial data on informative parameters, materials obtained during detailed exploration of the deposit and special engineering and geological surveys are used. On their basis, the variability of natural factors is studied, a model of their spatial location is adopted, and geomechanical maps and sections containing all the necessary information for calculating the stability of quarry slopes are constructed.

The relationship between the zoning indicator and informative parameters is described by the scales of points of influence of various factors on the stability of quarry slopes. The scoring scales are made on the basis of the dependences of the change in the reserve coefficient when changing the values of informative parameters, established by the results of mathematical modelling on the computer.

The sum of scores from all factors determining slope stability characterises the stability of the slope at given parameters and can be used to assess the stability indicators of slope height limit, slope angle of stable slope, stability coefficient.

Full scoring scales and examples of calculating zoning indicators have been published by the authors earlier [59].

A critical issue in zoning is the choice of one or more zoning criteria to divide the site into areas within which conditions are relatively similar.

When calculating stability using the method based on the theory of limit equilibrium, the slope is considered stable if the coefficient of

of stability is greater than 1, i.e. the sum of the forces shifting the slope is less than the sum of the forces holding it.

Thus, the stability coefficient equal to one should be taken as a boundary value when dividing the quarry face into stable and unstable sections.

The difficulty lies in the fact that the determination of the stability coefficient is associated with inevitable errors in the compilation of point scales (impossibility to enumerate all slope variants, errors of generalised regression equations), as well as with the influence of variability of rock mass properties.

In general terms, the mean square error in determining the stability factor is as follows:

$$m_n = \sqrt{m_c^2 + m_\delta^2} \tag{1.7}$$

where tc *is the* mean square deviation of the stability coefficient due to variability of rock properties; ms is the mean square deviation of the score.

The first component of the error mn was determined by simulation modelling of the stability factor distribution. For this purpose, the Monte Carlo method was used. Normally distributed values of density, adhesion and internal friction angle were generated on the computer and the stability coefficient of the same slope was repeatedly calculated.

It was assumed that the coefficients of variation of physical properties of rocks are as follows: density 6 %, cohesion 40 %, angle of internal friction 17 %. It turned out that the distribution of the stability coefficient obeys the normal law. The standard deviation of tc was ±0.075.

To estimate the value of ms, the estimated values of the stability coefficient were compared with the estimates of this coefficient obtained from the scoring scales. The obtained standard deviation is ±0.13.

Thus, the error value mn is equal to ±0.15.

If we set a confidence level of 0.95, the errors in determining the stability factor can exceed twice the value of mn in only five cases out of a hundred.

Consequently, the following boundary values of the stability coefficient can be accepted as a criterion for zoning of the quarry face by stability:

$n < 0.7$ - the board section is unstable;
$0.7 < n < 1.3$ is the zone of uncertainty;
$n > 1.3$ - the board section is stable.

The final stage of rezoning is the construction of a forecast map, on which areas of the board are allocated in accordance with the adopted rezoning criterion.

The predictive zoning map allows solving the following problems related to the stability of quarry slopes:

- Identification of areas of the board for which more accurate stability verification calculations are required (zone of uncertainty);
- Determination of observation station locations in potentially hazardous areas to organise instrumental observations of slope deformations (uncertainty zone, unstable areas);
- identification of unstable sections of the sidewall for the development of deformation control measures;
- Identification of stable sections of the slope where measures to improve production efficiency (increasing the height of the slope or increasing the slope angle) can be implemented.

Conclusions

As a result of the performed analysis, the following main conclusions can be drawn:

1. In this regard, the development of a reliable scoring system for assessing slope stability based on the theory of limit equilibrium and a methodology for zoning geomechanical conditions remain important research challenges.
2. It should be noted that the calculation of geomechanical rating classification RL does not take into account the type of drilling and blasting operations. The use of contour blasting will contribute to the correspondence of the real state of the near-contour part of the rock massif to the given results of the ratings calculation.
3. The analysis has established that the results of the assessment of the stability state of the instrument massifs depend on the minimum value of the stability reserve coefficient.
4. As a criterion of zoning was taken the degree of stability of the quarry sides by the value of the reserve coefficient, the total displacement of the surface of the instrument massifs, where the horizontal component of the shear vector is predominant.
5. The analysis has shown that the final stage of zoning is the construction of a predictive map, which allocates areas of the board in accordance with the adopted criterion of zoning, predictive map of zoning allows to solve problems associated with the stability of quarry slopes.

CHAPTER 2

STUDY OF THE CURRENT STATE OF GEOMECHANICAL CONDITIONS OF THE ROCK MASSIF FIELDS NORTH AND CENTRAL AMANTAITAU

The following items are included in the list of works to be performed during the study of the current state of geomechanical conditions of the rock massif:

- study mining and geological, hydrogeological and mining conditions of the Amantaytau deposit;
- analyses of previously performed works at the quarry to study the physical and mechanical properties of rocks and fracturing of the massif;
- estimation of horizontal stresses based on data of closely located analogue deposits and geodynamic zoning of subsurface;
- selection of drilling locations and parameters, additional engineering and geological boreholes.

It should be noted that during the collection of initial materials for the study of mining-geological, hydrogeological, engineering-geological conditions of the deposit the amount of useful information was insignificant and difficult to process. As the main source of data on these parameters, we received the report of sub-metallic exploration of the deposit [60], made by the State State Enterprise "Samarkandgeologia" in 1994, in which the information necessary for the subsequent assessment of the stability of the sides of the open pit in their design contours, is given literally a few pages of text. As for the characteristics of physical-mechanical properties of the rocks composing the deposit and the nature of their fracturing, they are presented as extremely approximate values with a very large spread, which does not provide the requirements of statistical validity and reliability, and for some types of rocks, which are very widespread, these data are not available at all. A more detailed description of the characteristics of the deposit rocks available in the stock literature will be presented below, in the relevant sections.

fracturing (shale) of the rocks composing the Central section, which will further allow to substantiate the forecast of the weakening surfaces development at the depth of the projected open pit mining.

We have developed and justified the list of additional works on based on the report of detailed exploration of the field, the number and depth of additional geotechnical wells, the order of their testing and recommendations on the technological regulations of drilling operations (drilling diameters, technology of penetration and testing by intervals, preliminary geological and technical columns of wells) Appendix 1,2.

Graphical materials are represented by the map, the actual material on which, in addition to the lines of the design calculation sections to assess the stability of the sides in the most relevant directions, are shown points of geological and structural survey Appendix 3. Features of the geological structure of the opened section of the Central Pit and drilling points of additional geotechnical and exploration wells, as well as previously drilled wells [60], used to build the calculation profiles.

The map of the actual material [60], which shows all drilled exploration wells (1160 wells), unfortunately, is characterised, first of all, by a minimum of information on the geological structure of the sides of the future open pit in its design contours of mining. The predominant part of the wells is concentrated in the central parts of both deposits. The common disadvantage of all prospecting and exploration works carried out in the 60-90s (even in the case of detailed exploration) is the absence of wells in the contours of the future slopes of the sides, beyond the boundaries of the productive strata. In principle, this is understandable: first of all, exploration works provided for calculation of mineral reserves, and the future need to assess the stability of the design contours of mining was regarded as something incidental and of little importance. As a rule, the slope angles of future open pits were set in the projects according to tabular recommendations [61], which in most cases did not cause special problems later on when developing deposits in rocky rocks.

Some of the missing information on the properties of contacts and other weakening surfaces in rocky and semi-rocky rocks of the Palaeozoic basement can be obtained from the scientific, production and regulatory literature [30-34], which summarises the results of studies on similar already developed and depleted fields.

Summarising the above, it should be noted that reliable and specific data on physical and mechanical characteristics of Meso-Cenozoic sediments (internal friction angles and cohesion of clayey rocks, marly clays and sands) can be considered practically absent at present; for rocky rocks only density values and values of resistance to uniaxial compression are available, and the scatter of limits of these values is such that it is practically impossible to use them. In addition, the determination of the strength characteristics of rocky and semiscale sedimentary and metamorphosed rocks requires the application of empirical formulas using tabular coefficients and the determination of the values of the structural weakening coefficients of the massif [62-65].

Drilled exploration wells are available only in analogue format, in the form of drilling documentation logs and geological and technical columns. There are no geological sections of potential design profiles to assess the stability of the pit sides along the most deformation-hazardous directions, which requires additional studies of the rock mass.

§ 2.1. Geological structure, structural-tectonic, hydrogeological, engineering-geological conditions and physico-geological mechanical characteristics of the rocks of the deposit.

The Central and Northern Amantaytau fields under study are located in Navoi region of the Republic of Uzbekistan.

The Auminzo-Amantai ore field, to the eastern end of which the Central deposit is located, is associated with the northeastern end of the Auminzatau mountain range, while the North deposit is located in the intermountain plain between the Auminzatau and Muruntau mountain ranges. The deposits are located between other large similar deposits-caries: Muruntau in the north-east and Daugyztau in the south-west.

The climate of the district is sharply continental. The average annual temperature is $+13.5^{0}C$. Summers are hot, dry, with average monthly temperatures of about $+29^{0}C$, with maximum temperatures in June - up to - $40{:}45^{0}C$.

Snow cover in winter is insignificant. The average annual precipitation does not exceed 100 mm, of which the main part falls in December-February and April-May. In spring there are heavy thunderstorm rains, which sometimes form short-term mudflows. The maximum daily precipitation reaches 10-12 mm.

Water resources of the district are limited. There are no surface watercourses, except for short-term ones during heavy rains.

In terms of geological structure and structural and tectonic conditions Amantaytau ore field is a series of linear ore-bearing structures that make up a single Amantaytau ore zone. Its thickness reaches 600-800 metres and its length is 2.8 km.

The Amantaytau ore field at the Central deposit is represented by Paleozoic sediments of the upper sub-formation of the Besapan Formation; at the Northern deposit it is overlain by Meso-Cenozoic sandy-silty-clayey sediments up to 150 m thick, in some places more.

Primary documentation preserved after drilling of exploration wells characterises the Paleozoic rocks mainly as mudstones, siltstones and, less frequently, sandstones. During the geological and structural survey on the sides and ledges of the advanced open pits of the Central deposit, we classified these rocks as clay (argillitic) shales for the most part, since the massif uncovered by the open pits in the predominant volume is unambiguously characterised by shale texture, which is shown in Figures 2.1 - 2.4.

Figure 2.1. East dipping 40^0 clay shale strata at the northern entrance to Pit 1A on the Central Section

Fig. 2.2. East dipping shale pack at angles 20-25^0 at the end of the southern working face of the Central pit, slope of the first ledge

Fig. 2.3. Fractured mudstones (siltstones) on the slope of the second scarp of the working western side of the Central pit

Figure 2.4. Argillitic clays in the slopes of the three upper escarpments of the northern working face of the Central Pit

The rocks described in the boreholes (siltstones, sandstones, mudstones), as a rule, have a massive texture and are less widespread. They are predominantly found on the ledges of the western side.

It is quite difficult to visualise the true structure of the massif under study from the fragmented core of low-strength sedimentary rocks extracted from small-diameter boreholes drilled with blowing, and only when the quarry is developed on its sides is an accurate and clear picture displayed.

However, when assessing the stability of the slopes, the type and name of the rocks composing them is of no fundamental importance: the main factors (in terms of geological features of the structure) determining the future stability of the slopes being built in rock formations are:

- nature and spatial orientation of fracturing (and in this case - shale) of the massif;
- physical and mechanical characteristics of the massif of these rocks, established along different directions of anisotropy.

The clays prevailing by thickness in the Meso-Cenozoic section, according to the visual assessment made during the geological survey of the northern working face, are characterised by low natural humidity (W0 ~ 10-15%), hard consistency (*J* fluidity index everywhere less than 0) and isotropic properties. Signs of layering are absent, the rock has a massive texture, rare fractures of separate fractures are jelly-ironed and break the massif into almost regular blocks-parallelepipeds up to 0.5-1.5 m in size. The described clays were discovered in the northern working face of the Central deposit, where three ledges are now built in them, starting from the day surface. The boundary of clay distribution on the day surface is shown on the actual material created during the geological and structural survey (Fig. 2.12.4). A characteristic feature of clayey rocks is their rapid soaking under the influence of precipitation. Spills from the slopes of the scarp slopes composed of clay are large (up to half a metre) blocks, some of which have completely lost their shape and have been blurred by the impact of water. Crumbling of slopes composed of clay soils is also characteristic.

According to drilling data of exploration wells, clayey marl interlayers with thickness from 15 to 30 m, in most parts - about 18-25 m, are found in the clay strata.

However, the most unfavourable fact in terms of ensuring the stability of the projected Northern pit sides is the almost universal presence of sufficiently thick sand horizons in the clay strata. The sand interlayers lie at rather great depths, which will create serious problems when building the sides: for example, from 55 to 85 m from the day surface on the western side of the future open pit (boreholes №№1652, 1745) with an average thickness of about 23-28 m (i.e. at least two ledges).

On the northern side (boreholes Nos. 1651, 1355), the depth of opening of the sand horizon in the clayey stratum was recorded as 61-96 m from the day surface with the dip of the sand roof to the north; the thickness of this horizon averages 24 m according to drilling data [60]. Thus, the dip of the sand horizon in the northern direction is traced here.

On the eastern side of the future quarry (see calculation profile No. 9, Appendix 1), borehole No. 1410 uncovered the sand horizon at a depth of 73 m from the surface, and its thickness was 44 m here. The total thickness of the Meso-Cenozoic clayey strata in this borehole was found to be 136 m.

While the Paleozoic rocks of the Paleozoic age, which lie practically from the day surface and are mined by the Central pit, are relatively well studied in terms of their physical and mechanical characteristics, the Meso-Cenozoic weak rocks and rocks have been studied rather poorly (especially in view of their importance and role in the condition of the sides and ledges of the projected pit). This factor necessitates additional drilling, sampling and laboratory testing at the North deposit.

The tectonic features of the Central and Northern deposits were studied in full at the stage of detailed exploration [60, P.90]; more specific information on the nature and spatial orientation of shale and fracturing of the rock massif was obtained during the geological and structural survey. All these data will be further taken into account in the development of 3D-models of the deposits and calculations of stability of the quarry sides (by means of drawing faults, fracture and shale fracture grid on the calculation profiles with the establishment of contact weakening surfaces).

Tectonic disturbances of the deposits are numerous and varied.

The main tectonic faults are the Central, Southern, Latitudinal, Intermediate, and Blocking faults, as well as faults hosting ore bodies.

During the geological and structural survey on the ledges of the Central pit sides, the orientation and character of weakening surfaces were assessed in 20 points, relatively evenly distributed over the area of the studied pit field. During the survey it was found that the main direction of fracture and shale separations dip is orientated mainly in the east and south-east directions. The predominant dip angles were 40-600, and to a lesser extent there are semi-dipping and steeply dipping fractures.

Hydrogeological and engineering-geological conditions of deposits are largely determined by climatic features of the desert region, where evaporation rate exceeds the amount of precipitation almost 20 times, and its confinement to a weak water-bearing stratum of strongly dislocated sediments (sandstones, siltstones, shales).

No water occurrences have been noted in Meso-Cenozoic sediments in the field area. There are only isolated cases of groundwater opening at the contact between Meso-Cenozoic and Palaeozoic sediments. This fact should be taken into account as an additional threat to the stability of the sides due to soaking of clay and rock contacts, accompanied by a decrease in the cohesion value at these contacts.

All lithological and petrographic differences of Palaeozoic rocks, due to their fracturing, form a single water-bearing complex. Groundwater accumulation is mainly associated with large discontinuities (Central, Southern, North-Eastern, etc.). The source of groundwater supply is atmospheric precipitation. Thus, the Amantaytau field is generally poorly watered. According to the degree of influence of groundwater on rock stability, it is categorised as complex due to the reduction of rock stability due to the presence of water in cracks, faults, and carbonified areas.

Physical and mechanical characteristics of rocks in terms of engineering and geological conditions Amantaytau deposit belongs to a complex type, and in seismic terms to a highly seismic area.

The upper Meso-Cenozoic strata, represented by hard clays with subordinate interlayers of marls, marly clays, loose and consolidated sands, is characterised by the following values of the main (determining the stability of sides and ledges) indicators of physical and mechanical properties according to the results of detailed exploration [60] Table 2.1.

Obviously, the information presented in Table 2.1 is insufficient for calculations of stability of the pit sides built in these rocks, while the height of these sides in them at the Northern deposit (according to the drilling results [60]) can reach 120-165 m. There is no information on the value of specific cohesion of clays (there are only values of internal friction angles). The strength characteristics of siltstones have not been determined, although their presence in the thickness of meso- Cenozoic sediments is not significant. The most qualitatively studied are marls, the interlayers of which are quite widely developed in the clayey strata.

Table 2.1.

Physical and mechanical characteristics of Meso-Cenozoic sediments of the Severnoye field

Name of breed	w_0,%	Volume-weight, t/m^3	Specific weight, t/m^3	Coupling,kPa	Angle of internal friction, deg.	Compressive strength, MPa	
						naturally occurring	water saturated
Hard clays	9,3 14,2	1,76-1,92	2,71-2,74	Not op.	12-19	Not op.	Not op.
Siltstones	Not op.	2,37-2,56	2,61-2,72	Not op.	Not op.	0,6-27,0	0,4-18,0
Mergels	Not op.	2,16-2,38	2,63-2,71	53	33	26,6	Not op.

The main disadvantage of the source materials for the upper layer of the Northern deposit, which excludes the possibility of a qualitative computational assessment of the stability of the sides and individual ledges, is the lack of any data on the strength and physical characteristics of the sands, which lie in a rather thick horizon in a relatively cohesive and strong clay stratum.

According to detailed exploration data, the Paleozoic complex of the Central and Northern deposits is composed of quartz-feldspar-mica sandstones (50%), carbonaceous-mica-quartz siltstones (35%), shales of the same composition (15%), strongly dislocated, oxidised and sulphidised. It should be noted that during the geological and structural survey of the Central Pit ledges, shale rocks were more widespread and sandstones were less widespread.

The values of the main physical parameters for different lithological varieties of rocks of the Paleozoic complex are given in Table 2.2.

Table 2.2.

Physical and mechanical characteristics of rocks of the Palaeozoic complex

Name	Volume weight, t/m^3	Compressive strength, MPa		Strength coefficient according to M.M. Protodyakonov
		naturally occurring	water saturated	
Sandstones	2,52 - 2,69	2,7 - 123,2	4,6 - 81,2	2,5 - 12,3
Siltstones	2,53 - 2,71	6,8 - 100,2	1,3 - 53,7	1,7 - 10,0

Slantsy	2,61 - 2,71	6,6 - 109,7	3,2 - 79,8	4,1 - 10,0

Obviously, to obtain complete information on the mechanical properties of rocks composing the deposits, it will be possible to use the tables presented in the relevant regulatory and research literature, where the necessary characteristics are determined for similar deposits in Uzbekistan and other regions on the basis of in-situ tests and back calculations [61-63]. This fully applies to the determination of friction and adhesion angles on rock contacts and on weakening surfaces formed by sliding surfaces, shale-schistosity separations, etc.

A significant lack of information presented in the detailed exploration report [60] for all rock types composing the North and Central deposits is the absence of results on the determination of such characteristics as lateral pressure coefficient, Poisson's ratio, deformation modulus, which are necessary for the evaluation of the stability of the sides by the finite element method. Such tests have not been carried out. Accordingly, these parameters should be determined with the necessary accuracy and in the numbers of determinations required for their statistical processing.

§ 2.2. Estimation of main horizontal stresses based on information about of closely located analogue fields and according to the geodynamic zoning

Geodynamic zoning includes assessment of the natural stress state of the rock massif, its tectonic structure. Identification of the block structure of the deposit, kinematic interaction of the blocks creating tectonically stressed zones, as well as determination of the stress field parameters will ensure safe mining conditions and more efficient development of the Amantaytau deposit.

According to the general displacements of tectonic blocks, the determining direction of the greatest principal stress σ1 in the field area is characterised by meridional orientation Fig. 2.5.

Regionally, the deposit is located in the same high grade structural block as the Muruntau deposit. The deposit is located at a small distance from Amantaitau, has similar genesis of formation and similar parameters of the stress field of tectonic nature. According to the results of high-precision geodetic observations of the processes of displacements of the instrumental massif at Muruntau, the largest displacements from the natural stress state were obtained in the submeridional direction [61]. That is, this fact confirms the direction of the greatest main stress in the meridional direction. The tectonic nature of the stress field is characterised by the horizontal location of two principal stresses, including the largest σ_1.

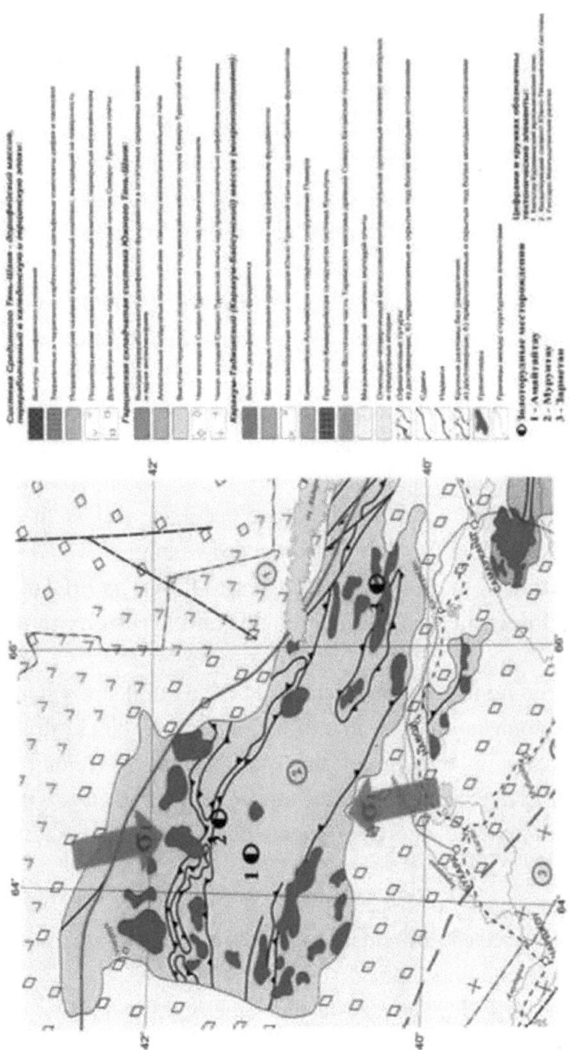

Fig. 2.5. Scheme of regional tectonic zoning

The geodynamic scheme of the area of deposits of similar type was investigated in detail, according to the stress state of the Kansai lead-zinc deposit. According to geodynamic zoning, the greatest main stress σ1 in the area of the deposit has submeridional direction and horizontal occurrence (Fig. 2.6).
[61]).

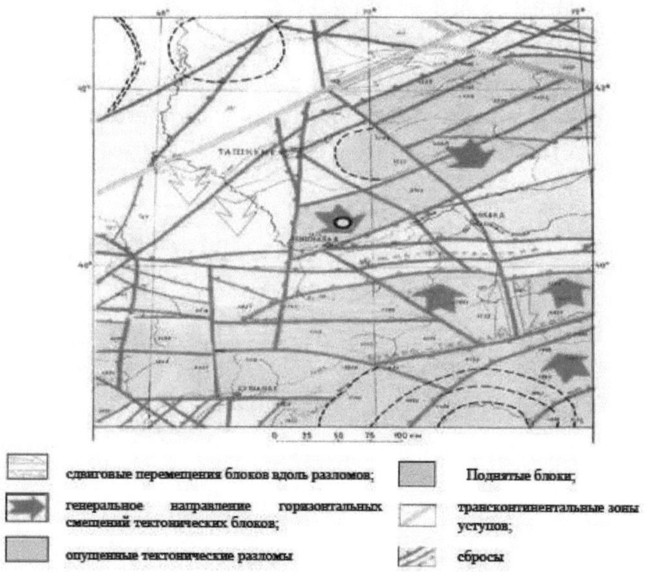

shear displacements of blocks along faults;
Raised blocks; general direction of horizontal displacements of tectonic blocks;
transcontinental scarp zones;
dipping tectonic faults dumps

Fig. 2.6. Geodynamic scheme of the deposit area similar type

As can be seen from the given scheme, in the whole area the greatest principal stress has a submeridional direction. The ratio of the main stress values for the Kansai field at a depth of 780 m is as follows:

$$\sigma_{верт} = 22МПа;\ \sigma_{гор.мерид.} = 42МПа;\ \sigma_{гор.широт.} = 32МПа;$$

at $\sigma_{верт} = \gamma \cdot H = 780 \cdot 0{,}028 = 22МПа;$

The formation and development of mountain fold structures in Central Asia is associated with horizontal compression and deformation of the Earth's crust in this region. Horizontal stresses exceed the vertical component. The sublatitudinal orientation of most megafolds corresponds to the submeridional direction of the main compression.

Numerous studies of the natural stress state of rock massifs in Central Asia have been carried out by specialists of the IFMHP of the Kyrgyz Academy of Sciences [62]. Analysing the results of natural stress determinations obtained at various mines, it was noted that the same-name horizontal stresses in rock massifs of deposits located at distances of hundreds of kilometres within Central Asia, at equal depths and in rocks close in strength, differ only slightly from each other. Statistical processing of the collected experimental material on natural stresses in rock massifs allowed to obtain average values of the main horizontal stresses, their change with depth at the value of the vertical component $\sigma_{верт.} = \gamma \cdot H$:

For strong rocks with modulus of elasticity E (5:6) 104 to (10:11) 104 MPa

$$\sigma_{гор.мерид.} = 5 + 1{,}86 \cdot \gamma \cdot H \text{ (МПа)}; \quad (2.1)$$

$$\sigma_{гор.широтн.} = 4{,}5 + 1{,}12 \cdot \gamma \cdot H \text{ (МПа)}. \quad (2.2)$$

For rocks of medium strength with modulus of elasticity

$$E = (2 \div 3) \cdot 10^4 \text{ до } (5 \div 6) \cdot 10^4 \text{ МПа}; \quad (2.3)$$

$$\sigma_{гор.мерид.} = 3 + 1{,}14 \cdot \gamma \cdot H \text{ (МПа)}; \quad (2.4)$$

$$\sigma_{гор.широтн.} = 2 + \gamma \cdot H \text{ (МПа)}. \quad (2.5)$$

where: H - depth of excavation, m; γ - density of overlying rocks $\gamma = 0.027$ MN/m^3.

Based on the average strength of Amantaytau rocks, we take the values of horizontal stresses of the second group for the deposit.

§ 2.3 Assessment of available materials and justification of the list of missing information to missing information for zoning of the quarry sides by stability

Based on the analysis of the available materials of the detailed exploration of the field on 12 April 2019, we established a list of wells previously drilled on the axes of the design profiles and necessary for the construction of geological sections along the axes of these profiles. The position of the design profiles is shown on the map of actual material and projected fieldwork volumes Fig. 2.7. The list of boreholes is given in Table 2.3.

Table 2.3.

List of boreholes required for construction of calculated stability profiles of the sides of the Central and Severny open-pit mines

Quarry (site)	Board	Calculation profile (Fig.1.)	Numbers of wells for construction of calculation profiles	Wells provided
Central	SOUTH-WEST	1	2140, 1568, 710, 421, 634, 633, 1571	1568, 710, 421, 634, 633, 1571
-* -	3	2	363, 521, 241	521, 241
North	SOUTH-WEST	3	1807,1975,1794,1806, 266, 292	1975,1806, 266, 292
-* -	SOUTH-WEST	4	1874, 1868, 1461, 1477, 1480, 1478	1874, 1461, 1477, 1480
-* -	3	5	2167, 1433, 1442, 2008, 1396, 2290	2167, 1433, 1442, 2008, 1396
-* -	N-W.	6	1420,1336, 1652, 1745	1420, 1336, 1652, 1745
-* -	C	7	1739, 1727, 1651, 1355	1651, 1355
-* -	S-B	8	1656,1340, 1633, 1634, 1757	1656, 1633, 1634
-* -	B	9	1766,1430, 1617, 1709, 1410	1430, 1617, 1410
-* -	U-W	10	2016, 1796, 1859, 1811	2016, 1811
Central	B	11	1362, 1564, 1980a, 1980, 1982, 1790, 151, 1508	1362, 1564, 1980a, 1980, 1982, 151, 1508
-* -	U-W	12	1556, 1971, 1943, 1517, 1563	1556,1517, 1563
TOTAL:			**63 wells.**	**45 well.**

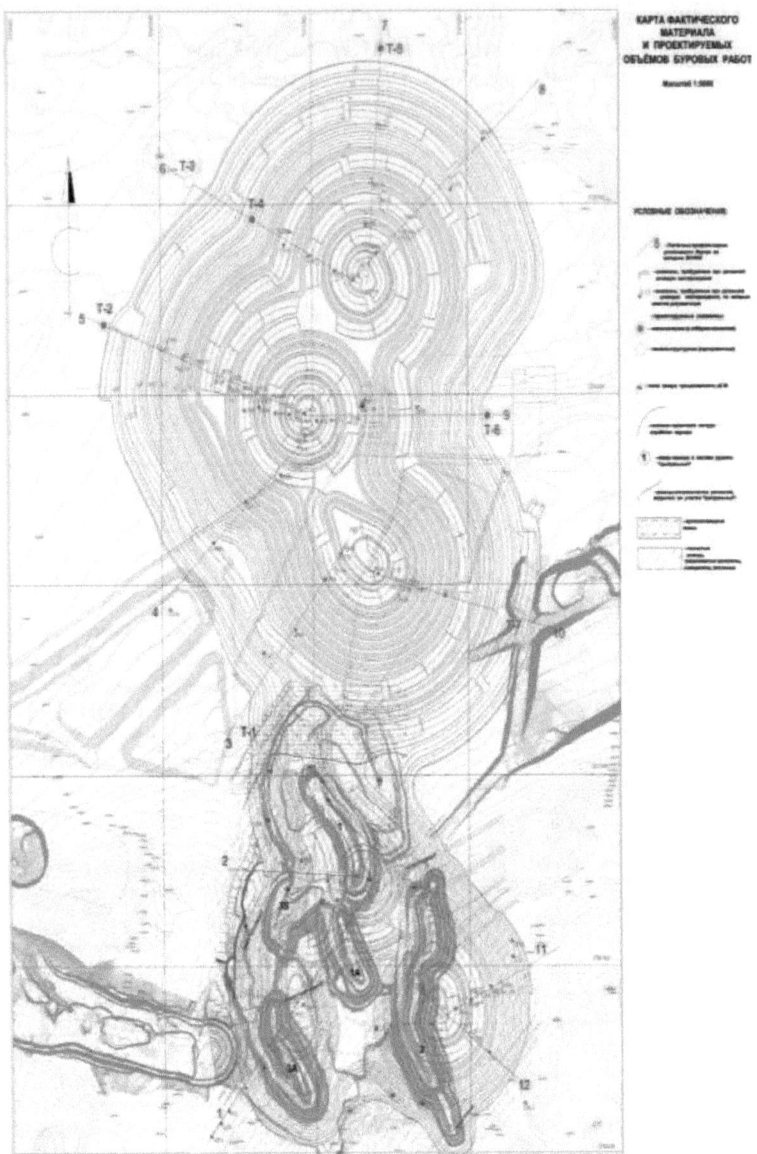

Fig.2.7. Position of design profiles on the map of actual material and projected volumes of field works

After careful analysis of all available materials, it was determined that for a full-fledged and qualitative assessment of the stability of the sides of the Central and Northern quarries in their design contours it is necessary to specify the geological structure in the most critical areas, precisely along the axes of the calculation profiles adopted by us. The boreholes presented in Table 2.3 partially allow to build these profiles and to establish the boundaries of various engineering and geological elements along them, as well as to draw the contact of Paleozoic rock and Mesozoic-Cenozoic clayey-sandy rocks, representing the weakest thickness of the geological section in terms of ensuring the stability of the sides in the design contours.

Analysing the map of the actual material (Fig. 2.7) it is easy to establish "white spots" on the future calculation profiles, namely, on their initial segments, characterising the geological structure of the instrumental strip, where the formation of the collapse prism is possible in the future. The available data on the previously drilled exploration wells (they are marked in green on the map), allow us to assess the geological structure of the profiles in fragments, but it is obvious that these profiles would become much more informative after using the columns of all the requested wells (their icons on the map are not filled with colour).

Due to the unavailability of several required wells, we have identified an additional list shown in Table 2 .4.

Analysing the map of the actual material (Fig. 2.7) it is easy to establish "white spots" on the future calculation profiles, namely, on their initial segments, characterising the geological structure of the instrumental strip, where the formation of the collapse prism is possible in the future. The available data on the previously drilled exploration wells (they are marked in green on the map), allow us to assess the geological structure of the profiles in fragments, but it is obvious that these profiles would become much more informative after using the columns of all the requested wells (their icons on the map are not filled with colour).

Table 2.4.

Additional list of exploratory wells required to for the construction of calculation profiles

№	No. of design profile	List of required wells
1	Profile 1	**662,709**
2	Profile 2	363 (reorder)
3	Profile 3	1384
4	Profile 4	**1441** (or 1873) It is desirable to find both
5	Profile 5	**1689, 1653, 1416 (or 2151), 1751 (or 2003), 1931, 1720, 2004,2194**, 1677, 2157, 1464, 1744, 2002, 1344, 1720
6	Profile 6	**1688**
7	Profile 7	**1332, 1661**
8	Profile 8	**1660**
9	Profile 9	**1409** (or SH-10 barrel documentation)
10	Profile 10	**1384, 2178, 1730** (or 1446), **1426**

11	Profile 11	**1790**
12	Profile 12	**1913** (or 2203), **1512** (or 1511)
Note: Bold font indicates wells that are required to be found		

The same map shows the points where additional geotechnical and exploration (mapping) boreholes should be drilled on the initial sections of the design profiles. The number of these boreholes is minimised to 7 for both projected open pits.

Geotechnical boreholes are designed for sampling of rocks occurring at rather large depths (about 100-150 m).

Exploration (mapping) boreholes are designed mainly to clarify the geological section in the area of the future instrumentation strip, to establish the position of the roof of strong Paleozoic rocks, the angle of dip of their layering (shale), to establish the spatial orientation of the contact between Meso-Cenozoic clayey soils and strong rocks.

In addition, the samples will be tested under laboratory conditions to establish the parameters of such parameters as deformation modulus, Poisson's ratio, etc., which are necessary for the evaluation of the stability of the sides using the finite element method.

Based on the above, we have formed a list of missing materials and information, which are necessary for zoning of quarry sides on stability.

To justify the optimal number of additional boreholes required and the order of their testing, let us refer to the map of actual material and projected volumes of drilling works (Fig.2.7). On it 12 calculation profiles are plotted, on which further calculations of stability of quarry sides in the most problematic areas will be made. The profiles are orientated along the normal to the berms, the distance between them corresponds to the requirements to the stability assessment of the quarry sides.

Let us consider the study of the section sides in the areas where these profiles are located.

- Profile 1 is located on the south-western curve of the Central Pit. It is illuminated by the available boreholes #421, 633, 634, and there is a fracture measurement point on the project slope of the rim. Additional surveys are not required here: the rim rounding is provided with information for stability assessment. For a more complete characterisation of the geological structure of the area and subsequent 3D-modelling it is necessary to obtain documentation for boreholes Nos. 709, 662 and 2140. The profile is completed by the fracture measurement point #18 at the downhole part of open pit No.1.
- Profile 2 characterises the western flank of the Central Pit and allows us

to assess the stability of the cantilever-shaped ledge on the project flank contour: such "ledges" usually have the potential to collapse. For this profile it is necessary to obtain information on well #363, which was drilled almost on the future instrumentation strip. The nature of fracture orientation here will be assessed using three points of the structural survey, and the lithological section will be assessed using the same points and wells #241, 521, the documentation of which has been provided to us.

- Profile 3 will provide stability calculations for the south-west roundabout of the North Pit. Based on drilling data, Meso-Cenozoic sediments are already widespread here. No information is available for boreholes #1734, 1384, 1807. Well No. 292 is the closest to the edge of the project side, further on the project apparatus strip is not studied at all. On this basis, it is planned to drill a mapping well No.T-1 at the initial section of the profile, the purpose of which is to establish the position of the rock roof beneath the Meso-Cenozoic clay and sandy rock interval, as well as the possible presence of groundwater.
- Profile 4 allows to perform stability calculations along the south-western side of the North pit. This profile is reasonably well studied for boreholes #1478, 1480, 1477, 1461, 1874. Documentation for boreholes #1141, 1873, 1868 should be provided for a more accurate cross section. No additional drilling is required here.
- Profile 5 characterises the stability of the western side of the North Pit, and information on its geological structure is extremely limited. The majority of the boreholes drilled along this profile (see Tables 2 .3-2.4) have not been found to date, although there are quite a few (see appendix graph). Nevertheless, it is necessary to drill geotechnical borehole No. T-2 in the area of the future instrumentation strip with sampling of undisturbed Meso-Cenozoic clays for further laboratory tests. The nearest well #1396 drilled to a depth of 220 m is used as an analogue of the section . The well was drilled in 1987, using the core drilling method, at an angle of 80^0 and an inclination azimuth of 270^0. According to its geotechnical section, the depth of the additional borehole No.T-2 is 130 m (up to penetration and testing of Paleozoic rocks, in this case - sandstones). In the interval of 11-21 m and 41-50.7 m interlayers of marl are revealed; from 50.7 to 80.4 m there are sands of uncertain density. Accordingly, clay monoliths should be sampled from the depths of 10, 25, 30, 40 m (the total number will be 4 pieces), marl is sampled from the depths of 15, 45 m (2 monoliths). A special requirement is to sample the sandy horizon with the maximum possible number of samples (at least 6 samples from the interval of 50-80 m).

The last sample is taken from sandstones from the bottom of the well. The approximate total number of samples from well T-2 is 13.
Drilling of geotechnical borehole T-2 should be carried out by core drilling method with a diameter of not less than 112 mm (to obtain quality samples), with compressed air blowing.
Besides, in case of absence of documentation of boreholes Nos. 1689, 1416, 1653, 1677, 2157, it will be necessary to drill another mapping borehole on this profile, in the slope zone of the projected side (see appendix graph); the depth of this borehole should be the same (130 m), sampling is not necessary, the main condition is to uncover Paleozoic rocks.

- Profile 6 on the north-western curvature of the North pit face is the least informative: the side strip and the upper ledges of the face up to the middle of the project slope are completely unexplored. The downhole part of the future open pit is characterised by wells Nos. 1420, 1336, 1652, 1745. The documentation for well No. 1688 has not been found.

In this regard, two boreholes should be drilled at the side section of the profile: geotechnical borehole No. TT-4 and mapping borehole No. TT-3. An analogue of the section of these boreholes is taken from borehole #1745, drilled 120 m from borehole #T-4 and 250 m from borehole #T-3.
The thickness of the Meso-Cenozoic strata penetrated by borehole No. 1745 is 134.0 m; in the interval of 56-84 m, a horizon of loose sands is exposed, which should be sampled.
Well #T-4 will be sufficient to drill to the opening and passage of the sand horizon (to a depth of 90 m). At the same time monoliths of undisturbed clay with marl interlayers from depths of 20, 40, 50 m (3 monoliths) and sands with the maximum possible number of samples (minimum 6 from 28-30 -m interval starting from the depth of 60 m) should be tested. Drilling should be carried out by core drilling, with a diameter of at least 112 mm.
Well #T-3 will need to be drilled to establish the position of the loose rock-rock contact and the presence of an aquifer to a depth of 140 metres. No sampling will be carried out here and drilling is possible using a pneumatic hammer, compressed air, or dry-hole drilling with blowing.

- Profile 7 runs normal to the north face of the North Pit. In this area, the ridge is also unexplored, and the nearest well #1355 (drilled in 1985 to a depth of 508 m) is 165 m horizontally distant from the project ridge. In addition, it is necessary to find documentation on the drilling of wells #1332 and 1661.

On this profile it is planned to drill one geotechnical borehole No.T-5 (see graph. appendix), the expected section of which in comparison with the geological column of borehole No.1355 looks as follows: yellowish, grey solid clays lie to a depth of 187 m and are underlain by Paleozoic sandstones; in the interval of 8097 m marls are uncovered; from 97 to 120 m a horizon of fine-grained sandstones of Meso-Cenozoic age is identified.

- The depth of the additional geotechnical borehole #T-5 is set at 120 m; this depth will be sufficient to sample the clayey strata with monoliths at 20 m sampling intervals (4 monoliths), to sample 2 monoliths of marl (to compare physical characteristics with the data given in the Detailed Exploration Report [60]), and to sample sandstones to a depth of 120 m at 20 m intervals (2 samples).

The borehole should be drilled using a core hole with a diameter of at least 112 mm with air blowing. The lower boundary of the potential collapse prism here will be in clays due to their significant thickness, and it is not necessary to penetrate the Paleozoic rocks in this case.

- Profile 8 has been drilled in a sufficiently informative manner during detailed exploration, and no additional drilling is required here. A computational section for stability assessment can be constructed using the existing wells #1633, 1634, 1656, 1420, supplemented by the missing wells #1160, 1340.

- The profile 9 outlined on the eastern side of the North Quarry is critical for assessing the stability of the cantilever-type ledge on the project plan. To build the geological section of this profile: - existing boreholes Nos. 1410, 1617, 1430 are used;

- It is necessary to provide documentation for boreholes Nos. 1709 and 1409 (or documentation of shaft sinking of the ShKh-10 mine shaft) in order to establish the position of the contact between Meso-Cenozoic and Palaeozoic rocks more precisely.

In addition, in order to establish the thickness of clayey rocks on the instrument strip and to test them on the eastern side of the pit, geotechnical borehole #T-6 should be drilled to a depth of 90 m, which will be sufficient in this case to establish the position of the potential slip surface in the massif of the side.

Well No. 1410 drilled in 1985 at a distance of 230 m from the future edge of the rim to a depth of 385 m is used as an analogue for the approximate type of geological section of the rim at the drilling point of well T-6. According to the information available on it, the section of the projected borehole No. T-6 is expected to reveal hard argillite-like mottled clays with marl and dense sandstone interbeds.

This borehole should be drilled using a core drilling method, with monoliths taken at 10-20 m intervals (5 monoliths in total).

- Profile 10 is located on the south-eastern curve of the North Pit, in the area of the external dump. Only two previously drilled boreholes (Nos. 1811, 2016) are documented here, the first of which is located 200 metres from the project top of the pit. Drilling documentation for several more boreholes is required to complete the section, as listed in Tables 2.3 - 2.4.

Drilling of an additional mapping well, necessary to establish the depth of contact between Mesozoic and Cenozoic rocks, should be carried out to a depth of 50 metres. In borehole No. 1811, located 180 m away, the depth of penetration of the Paleozoic siltstone roof underlying the Mesozoic-Cenozoic clay strata was 45 m.

- Profile 11, which will be used to assess the stability of the eastern side of the Central Pit, should be cross-sectioned from the existing boreholes Nos. 1362, 1564, 1980a, 1982, 151, 1508, which were drilled along the entire calculation profile. It is necessary to refine the section for the missing well #1790. Additional drilling is not required here, as the boreholes have penetrated the Palaeozoic rocks practically from the day surface.
- Profile 12, laid out on the south-eastern curve of the Central Pit, as well as profile 11, has been sufficiently studied during detailed exploration of the deposit. Paleozoic rocks (siltstones, sandstones, shales) lie here from the surface and the profile is closed by well No. 1563. To build this profile, it is necessary to submit documentation on drilling of wells, the list of which is given in Tables 2.3 - 2.4.

Conclusions

1. Collection, systematisation and digitisation of exploration drilling data, detailed exploration of the deposit showed that there are no geological sections on potential calculation profiles to assess the stability of pit sides along the most deformation-hazardous directions, which requires additional studies of the rock mass.
2. A significant lack of information on the geological structure of the deposit is the absence of results on the determination of such characteristics as lateral pressure coefficient, Poisson's ratio, deformation modulus, which are necessary for the assessment of the stability of the sides by the finite element method. Accordingly, it is necessary to determine these parameters with the necessary accuracy and in the number of determinations required for their statistical processing.
3. Analysing numerous studies of the natural stress state of rock massifs, it was noted that the same horizontal stresses in rock massifs of deposits, within the limits of Central Asia, at equal depths and in rocks close in strength

allows to obtain the average values of the main horizontal stresses, their change with depth at the value of the vertical component $\sigma_{,41\iota} = \gamma\text{-}H$. According to which Amantaytau belongs to the second group by the values of horizontal stresses.

4. According to the analysis of available materials of detailed exploration we have established a list of additional boreholes for construction of geological sections for zoning of the sides for full and qualitative assessment of stability of the sides of the Amantaitau open pit in their design contours, exactly along the axes of the design profiles accepted by us.

CHAPTER 3

LABORATORY RESEARCH TO DETERMINE ENGINEERING AND GEOLOGICAL CHARACTERISTICS OF THE ROCKS OF THE ROCKS OF THE DEPOSIT

On the proposed method of selection and preparation of rock samples for laboratory tests of physical and mechanical properties of rocks and the method of research of physical and mechanical properties of rocks, on the determination of tensile strength, deformation properties, cohesion and angle of internal friction, on the velocities of elastic longitudinal and transverse waves in rocks.

§ 3.1. Methodology of selection and preparation of rock samples for laboratory testing of physical and mechanical properties of rocks laboratory tests of physical and mechanical properties of rocks. rocks.

Sampling (both from wells and from the slopes of the sides) and their packaging should comply with the requirements of GOST 12071-2014 [67], with the exception of the possibility of drying or artificial water saturation, not provided by the programme of laboratory tests. Samples shall be provided with well-readable labels, oriented according to the scheme "top-bottom" and wrapped in several layers of stretch film or packed in gauze soaked in melted paraffin with bitumen.

The methodology and quantitative volumes of laboratory tests on selected samples are not regulated, as these works depend on specific conditions. The main condition is the sufficiency of the number of tests on all selected samples for qualitative statistical processing according to the requirements of GOST 20522-2012 [68]. The use of tabulated values of physical and mechanical characteristics of rocks, given in the current regulatory literature, is not excluded.

56 samples were taken to carry out a set of tests to determine the physical and mechanical properties of rocks from the Northern and Central Amantaytau deposits.

Preparation of rock samples for laboratory tests included operations on depressurisation of samples (extraction from polyethylene bags), cleaning of their surfaces, manufacturing of rectangular samples of required dimensions by processing on special rock-cutting equipment and stacking of manufactured samples in exicators for storage.

The following equipment was used to prepare the specimens for testing:

- FUBAG A-44/420 M3F stone cutting machine ;

- POLYLAB P12M grinding machine;
- exciters.

The FUBAG stone cutting machine was used for high-performance cutting of rock into rectangular specimens with flat faces. The cutting was carried out with a diamond cutting wheel with a diameter of 350 mm Fig. 3.1.

Fig.3.1 FUBAG A-44/420 M3F stone cutting machine

Polylab P12M grinding machine was used for grinding the surfaces of rock samples after cutting them on the stone cutting machine Fig.3.2.

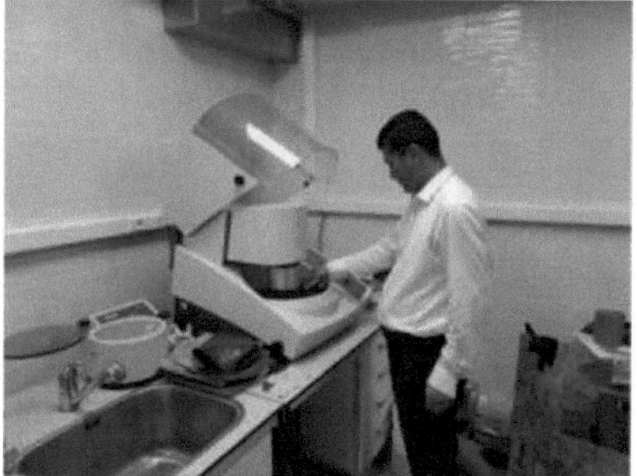

Fig.3.2 POLYLAB P12M grinding machine

Exicators, which are glass vessels with hermetically sealed lids, were used to store the manufactured specimens before testing without additional loss of moisture.

The dimensions of the manufactured rock specimens were chosen taking into account the requirements of GOST standards [69-78] for test methods and the presence of weakening surfaces, diverse in composition and orientation.

If it was not possible to produce specimens of the required height due to splitting along the weakening surfaces during preparation, specimens of the maximum possible height were tested.

§ 3.2. Methods of research of physical-mechanical properties rocks

Laboratory studies of rocks, as well as desktop processing of test results, were carried out in accordance with the requirements of the current regulatory documents, the main ones being:

GOST 25100-2011. Rocks. Classification, GOST 12071-2014. Selection, packing, transport and storage of samples, GOST 5180-2015. Methods of laboratory determination of physical characteristics, GOST 205222012. Rocks. Methods of static processing of test results.

Laboratory equipment used for rock research has passed metrological verification in accordance with the established procedure. In laboratory conditions, tests were carried out to determine physical and mechanical characteristics for the following types of studies:

- natural moisture content; rock density; dry rock density; moisture at the yield boundary; moisture at the rolling boundary; granulometric composition; water saturation coefficient; plasticity number; cohesion and angle of internal friction of rocks; cohesion and angle of internal friction by single plane shear method in natural and soaked states.
- compressive strength; tensile strength; shear strength; natural slope angle; calculation of internal friction angle and cohesion.

Density and humidity of rocks were determined in accordance with GOST 5180-2015.

The moisture content of clayey rocks was determined as the ratio of the mass of water removed from the rock by drying to a constant mass to the mass of the dried rock:

$$W = \frac{m_1 - m_0}{m_0 - m} 100 \tag{3.1}$$

where t - weight of empty bunker, g ; t_1 - weight of wet rock with bunker, g; t_0 - weight of dried rock with bunker, g, dried at temperature 105±20C.

Density was determined by the cutting ring method. The density of the rock is determined by the ratio of the weight of the sample to its volume in the natural state and is expressed in g/cm3

$$\rho = \frac{m_1 - m_0 - m_2}{V} \tag{3.2}$$

where m_1 - mass of the sample with the ring and plates, g ; t_0 - mass of the ring,

g ; m_2 - mass of the plates, g ; V - internal volume of the ring, $cm3$.

Rock density was determined by direct measurement and by weighing in water:

1. The density (volumetric) was determined by direct measurement. For this purpose, cylindrical samples of regular shape were measured and weighed, after which the density value was calculated. GOST *8269.0-97* p.4.1.3.

2. The density of rock samples (ρ, kg/m^3*)* was calculated according to the formula:

$$\rho = \frac{m}{V} \qquad (3.3)$$

where t is the mass of rock in some volume V.

3. Determination of soil density by weighing in water. The soil sample is covered with paraffin coating, immersing it for *2-3* seconds in heated paraffin. The cooled paraffinised sample is weighed in a vessel with water.

4. Soil density ρ, g/cm^3 is calculated by the formula

$$\rho = \frac{m\rho_p\rho_w}{\rho_p(m_1 - m_2) - \rho_w(m_1 - m)} \qquad (3.4)$$

where t - mass of soil sample before paraffinisation, g ; $t1$ - mass of paraffinised soil sample, g ; $m2$ - result of weighing the sample in water, difference of masses of paraffinised sample and water displaced by it, g ; ρ_p - density of paraffin, taken as 0,900 *g/cm3* ; pw - density of water at test temperature, *g/cm3*.

Determination of the moisture content of the yielding limit (WL) and rolling limit (WP) of clayey rocks.

The yield point WL should be defined as the moisture content of the paste prepared from the sample under test, at which the balance cone is immersed under its own weight for 5 sec to a depth of 10 mm.

The limit of WP rolling (plasticity) should be defined as the moisture content of the paste prepared from the sample under test, at which the paste, rolled into a bundle with a diameter of 3 mm, begins to disintegrate into pieces of length 3-10 mm.

The plasticity number IP of a clayey rock sample is determined by the moisture content difference corresponding to two rock states: at the yield boundary WL and at the rolling boundary WP.

§ 3.3. Determination of tensile strength, deformation properties, adhesion and angle of internal friction, Young's modulus of elasticity and Poisson's coefficient of Young's elasticity and Poisson's ratio by the velocities of passage elastic longitudinal and transverse waves in rocks.

Tensile strength and deformation characteristics of rocks are performed in accordance with GOST 24941-81 Rocks. Methods of determination of mechanical properties by loading with spherical indenters [73].

For the test we used a Phenom probe, Fig. 3.3. with a manual drive designed to determine the complex determination of strength and deformation characteristics of rocks in laboratory and field conditions on samples of arbitrary shape, including irregular.

The test mode provides two-stage loading and unloading of specimens with recording of corresponding changes in the dynamometer readings and indent convergence measurement indicators.

The value of the load corresponding to each loading stage depends on the rock strength, which is determined by the tabular data in the Phenom probe's operating instructions.

Fig.3.3. Phenom probe is designed for complex determination of strength and deformation characteristics of rocks

Tensile strength is calculated according to the formula:

$$\sigma_{p'} = 7{,}5\frac{P}{S} \qquad (3.4)$$

where P - maximum destructive load, *kg's; S* - value of the actual surface area of the through *fracture* (split) of the specimen F, *m2.*

Calculate the residual strain modulus (D) of the rock during indentation of spherical indenters (*kg^f/mm2*) according to the formula:

$$D = \frac{P_2 - P_1}{\delta_{2,0} - \delta_{1,0}} r_0 \qquad (3.5)$$

where r_o= 7.5 *mm* - indenters radius; $0_{2,o}$, $\delta_{1,o}$ - readings of indenters convergence measurement indicators during unloading, *mm.*

Calculate the modulus of elasticity in compression E from the results of measurements at the second stage of loading according to the formula:

where a_2 is the radius of the contact site, *mm* ; Δ_{2u} is the elastic approach to the sample, μm; P_2 is the load, *kg's.*

Rock shear resistance is defined as the ultimate mean tangential stress at which a rock sample shears through a fixed density at a given normal stress [74].

Shearing tests were carried out on universal machines of LFM-50 series, on samples of undisturbed structure, at natural humidity, without their preliminary compaction according to the scheme of unconsolidated shearing Fig.3.4.

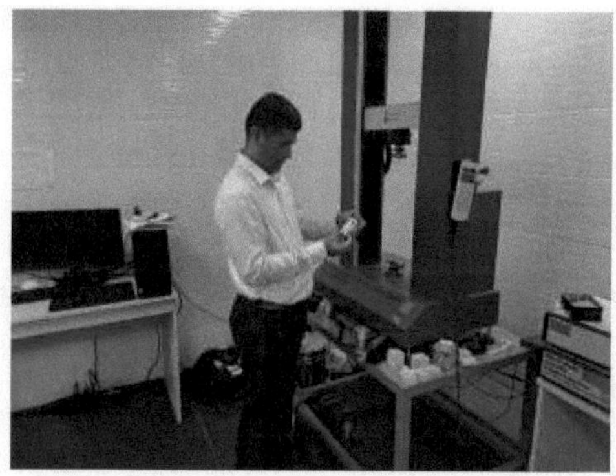

Fig.3.4 Shear test apparatus for dispersed rocks universal machine LFM-50 series

Tangential load was applied uniformly, at a rate of 0.05 mm/min until instantaneous shearing (failure).

Tangential and normal stresses τ and σ, MPa were calculated from the values of tangential and normal loads measured during the tests using the formulas:

$$\tau = \frac{Q}{F}; \qquad (3.7)$$

$$\sigma = \frac{F}{A} \qquad (3.8)$$

where Q and F - respectively, tangential and normal forces to the shear plane, kN; A - shear area, cm^2.

The angle of internal friction φ and specific adhesion C are defined as parameters of linear dependence:

$$\tau = \sigma \cdot tg\varphi + C, \qquad (3.9)$$

where τ and σ are determined by formulae (3.7) and (3.8).

Definition of Soakability. Soakability is the ability of rocks, when interacting with calm water, to lose cohesion and turn into a loose mass with partial or complete loss of bearing capacity. Soakability indicators, along with shrinkage, swelling, softening, characterise the water resistance of rocks, i.e. the ability of rocks to retain mechanical strength and stability when interacting with water.

Both rocky sedimentary rocks with soluble (carbonate) or clayey cement and various cohesive clayey rocks are susceptible to soakability. Disturbance of the natural composition of rocks contributes significantly to soakability. Most igneous, metamorphic, sedimentary rocks with crystallisation ionic-covalent bonds are practically non-wetting.

Soakability is caused by weakening and sometimes destruction of structural bonds between elementary particles or aggregates of rock in the process of its hydration.

The following soakability indices are used to characterise the soakability of rocks:

а) Soaking time - time during which a rock sample placed in water loses its cohesion and disintegrates into structural elements of different sizes;

б) Soaking rate - estimated by relative mass loss over time;

в) degree of soaking - characterises the relative reduction in height of the sample;

г) character of soaking - reflects the qualitative picture of rock sample disintegration and is established visually (rapid disintegration, flaking of separate crusts, disintegration into separate lumps, etc.).

According to RSN 51-84 [79], according to the time of sample soaking, differentiate types of soakability: instantaneous - completely in 1 minute; very fast - more than 80-90% in 30 minutes; fast - more than 50% in 1 hour; slow - less than 50% in 6 hours; very slow - less than 25% in 24 hours; not soaking rock - less than 10% in 48 hours.

In laboratory conditions, the rock samples were tested for soakability. A sample was cut from the core sample, placed on a sieve with a 10x10 mm mesh, suspended on a scale and immersed in a container with tap water at room temperature. During the process of soaking, the character and speed of soaking of samples were observed and recorded in logs.

The filtration coefficient of clayey and sandy rocks was determined on undisturbed samples using the compression device of the field laboratory PPL-9 [80].

The filtration coefficient expresses the water filtration rate at a head gradient equal to one and a linear filtration law

$$K = 0{,}01565\frac{Б}{t \cdot r} \tag{3.10}$$

$$Б = \ln\left(1 - \frac{y}{H}\right) \tag{3.11}$$

where K - filtration coefficient, *cm/sec* ; t - experiment duration, *sec* ; r - correction for water temperature (according to Khazin); y - water level drop in the pressure tube, *cm* ; H - initial head height, *cm.*

Dynamic methods of determining deformation (elastic) properties of rocks are based on measuring the velocities of elastic vibrations excited in the studied samples in the range of sound and ultrasonic frequencies, i.e. they are actually at the same time methods of determining the acoustic properties of rocks.

The most widespread in practice in the study of rock properties is the pulse dynamic method, which is based on passing through the sample of the rock under study repetitive pulses of ultrasonic oscillations, the values of propagation velocities of which are used to calculate elastic characteristics. This method was developed much later than the static method, but it is becoming more and more widespread due to its simplicity, low labour intensity of measurements and the use of easy-to-use and reliable serial measuring devices Fig.3.5.

The essence of the method is the measurement of the elastic pulse travelling time through a rock sample [78].

To carry out the test, the device for determining the velocity of elastic acoustic waves "Ultrasound" was used.

Each sample is sounded three times in mutually perpendicular directions.

Fig.3.5. The device for determining the velocity of elastic acoustic waves "Ultrasound"

Dynamic modulus of elasticity (Young's modulus) is determined by formula (3.12) "Properties of rocks and methods of their determination" [81]:

$$E_Д = \frac{C_{np} \cdot \rho \cdot (1+\mu) \cdot (1-2\mu)}{1-\mu}, \qquad (3.12)$$

where C_{pr} is the longitudinal wave velocity, *cm/sec* ; ρ is the density, $kg^\wedge sec^2/cm^4$; μ is the Poisson's ratio.

To determine the dynamic Poisson's ratio, the dependence according to the formula (3.13) "Properties of rocks and methods of their determination" is used [81]:

$$\mu = 0{,}5 - \sqrt{R} - R^2 \qquad (3.13)$$

where R is the ratio of the transverse wave velocity to the longitudinal wave velocity.

The ultimate strength of rock rocks in uniaxial compression, σ_{cj}, was determined in accordance with GOST 21153.2-84 [71].

The essence of the method is to measure the maximum breaking force applied to the ends of a regular shaped specimen through steel flat plates Fig.3.6.

Fig.3.6. Testing machine (laboratory press 50 t

Uniaxial compression tests were carried out on prismatic specimens (with a square cross-section), with the ratio of height h to the square side, approximately equal to 2.0-2.5, which at the side of the specimens 41±1 mm corresponded to their height equal to 80-100 mm. The end surfaces were parallel to each other and perpendicular to the side surface.

If it was not possible to produce specimens of the required height (due to splitting along the weakening surfaces during preparation), the specimens of the maximum possible height were subjected to compression tests.

According to the maximum load and cross-sectional area, the compressive strength of the specimen in MPa was calculated by the formula

$$\sigma_{сж} = K_в \frac{P}{S} 10 \quad (3.14)$$

,

where *P* is the force destroying the specimen, *kN;* S is the average cross-sectional area of the specimen, equal to the half sum of the areas of the upper and lower end of the specimen before its destruction, *cm*^{2}. Kv - dimensionless coefficient of specimen height, equal to 1.00 at the ratio of height to diameter *m = 2±0.05.*

Determination of uniaxial tensile strength of rock samples by the method of spherical indenters is performed in accordance with GOST 21153.3-85 "Rocks. Methods of determination of uniaxial tensile strength" [72]. [72].

Uniaxial tensile tests were performed on prism-shaped specimens with a height of 15-20 mm. The end surfaces were parallel to each other and perpendicular to the side surface. If it was impossible to produce specimens of the required height (due to splitting along the weakening surfaces during preparation), tensile tests were performed on semi-regularly shaped specimens allowed by GOST tensile tests [72].

The uniaxial tensile strength (σ_p) in MPa for each specimen is calculated according to the formula

$$\sigma_{p'} = 7{,}5 \frac{P}{S} \quad (3.15)$$

where *P* - destructive force, *kN; S* - surface area of the sample fracture, *cm*^{2}*;* *K* - dimensionless scaling factor, taken equal to 1.00, at *S = (15±3) cm*^{2}.

For other values of the sample fracture surface area S, the coefficient *K* is set according to Table 3.2 [72].

The compressive shear strength of rocks (normal and shear components of the load) is determined by the oblique shear method in accordance with GOST 21153.5-88 [77].

The method of oblique rock shear provides shear failure of samples on an inclined plane, with a given ratio of shear and normal components of the load, i.e. shear failure with compression and calculated by the formulas:

$$\tau_{\Theta} = 10\frac{P}{S}\cos\Theta \quad (3.16)$$

$$\sigma_{\Theta} = 10\frac{P}{S}\sin\Theta \quad (3.17)$$

where *P is the* breaking force, *kN; Θ is the* angle between the shearing plane and the direction of the breaking force, *deg ; S h=* ▪ d is the area of the shearing plane of the specimen, cm^2.

Matrices with different tilt angles (25, 35 and 45 degrees) Fig.3.7 were used in the test.

Fig.3.7. Laboratory press in the variant with interchangeable detachable dies at shearing with compression at an inclination angle of 450.

The methodology for determining the strength properties of rocks in water-saturated state is as follows: the samples were dried to constant mass in a desiccator at a temperature of 1050 C.

The dried samples were weighed and placed in a vessel of distilled water, immersing them in water to their full height.

The samples were kept for 6 days. Before weighing, the end faces of water-saturated samples were wiped with wet squeezed gauze.

The specimen was placed on the test equipment and then a compression test was carried out using the standard techniques given above.

§ 3.4 Results of studies of physical and mechanical characteristics of rocks of Amantaytau deposit

Laboratory studies of physical and mechanical properties of rocks were carried out on 56 samples taken from the Northern and Central Amantaytau deposits.

The obtained results of the rock properties study are presented in the tables.

1. The results of physical and mechanical properties of clayey rocks of Northern and Central Amantaytau deposits are given in Table 3.1.
2. The results of determinations of physical and mechanical properties of rocky rocks of the Northern and Central Amantaytau deposits are shown in Table 3.2.
3. The results of determinations of physical and mechanical properties of

loose rocks of the Northern and Central Amantaytau deposits are shown in Table 3.3.

Laboratory studies on determination of engineering and geological characteristics of rocks of the deposit were carried out in the "Innovative laboratory of diagnostics of structure and properties of rocks with the use of laser-ultrasound diagnostics" of the National Research Technological University "MISIS" (Russian Federation, Moscow).

Table 3.1.

Results of physical and mechanical properties of clayey rocks of the Northern and Central Amantaytau deposits

№ пробы	№ скважины, точки наблюдения	Глубина (интервал), горизонт отбора пробы, м	Кол-во образцов	Литотип	Влажность, %			Число пластичности	Показатель текучести	Плотность, г/см³			Угол внутреннего трения образца	Удельное сцепление грунта, МПа	Предел прочности при сжатии, МПа		Модуль деформации, МПа	Коэффициент Пуассона	Модуль упругости, МПа	Размокаемость	Коэффициент фильтрации
					природная	на границе текучести	на границе раскатывания			частиц	природной влажности	сухого			в крест простирания	по слоистости					
					W	W_L	W_P	J_P	J_L	ρ_s	ρ	ρ_d	φ	C	$\sigma_{сж}$	$\sigma_{сж}$	Ед	μ	E		Кф
1	2	3	4	5	6	7	8	9	10	11	12	13	14	15	16	17	18	19	20	21	22
1	т.н. 9	350	1	Глина твердая прочная тонкослоистая	7,62	53,76	22,11	31,65	-0,46	2,74	2,02	1,88									
1	т.н. 9	350	2	Глина твердая прочная тонкослоистая	12,07					2,74	2,07	1,85			4,07 - 5,96 / 5,22						
1	т.н. 9	350	3	Глина твердая прочная тонкослоистая	8,81					2,74	1,99	1,83				1,80 - 4,79 / 3,00					
2	т.н. 10	350	1	Глина твердая прочная тонкослоистая	7,98	59,22	28,75	30,47	-0,68	2,75	2,02	1,87					2,75	0,39	2,74		
9	Скв - 8	15,0	1	Глина твердая	30,17	39,59	18,48	21,11	0,55	2,74	1,91	1,47	21°48'	0,100		0,87 - 1,06 / 0,96	2,53	0,42	0,88	медленная	0,008
9	Скв - 8	15,0	2	Глина твердая	19,93					2,74	1,99	1,66	21°48'	0,105							
24	Скв - 6	17,0	1	Глина твердая	29,57	53,70	22,21	31,49	0,23	2,75	1,85	1,43	19°18'	0,06		0,11 - 0,39 / 0,25					
25	Скв - 6	18,0	1	Глина твердая	35,51	54,01	24,01	30,00	0,38	2,75	1,77	1,31	18°	0,09		0,19 - 0,25 / 0,22					
36	Скв - 3	30,0	1	Суглинок песчанистый плотный	19,65	46,94	30,06	16,87	-0,62	2,73	1,99	1,66	14°06'	0,018							
36	Скв - 3	30,5	2	Суглинок песчанистый плотный	20,04	47,22	30,43	16,80	-0,62	2,73	2,01	1,67					2,81	0,63	0,67		
41	Скв - 3	100,0	1	Глина насыщенная водой, полутвердая до пластичной	22,08	41,08	18,28	22,81	0,17	2,74	1,98	1,62	19°18'	0,115							
41	Скв - 3	103,0	2		21,94	47,91	20,92	26,99	0,04	2,74	2,02	1,66					2,53	0,44	0,97		
43	Скв - 2	25,0	1	Глина твердая	20,13	61,96	29,71	32,24	-0,30	2,75	2,22	1,85	20°30'	0,07							
43	Скв - 2	25,0	2	Глина твердая	18,75					2,75	2,20	1,85					2,77	0,45	0,93		
43	Скв - 2	25,0	3	Глина твердая	19,97					2,75	2,23	1,86								очень медленная	
45	Скв - 2	50,0	1	Глина черная твердая тонкослоистая	12,19																
45	Скв - 2	50,0	2	Глина черная твердая тонкослоистая	24,85					2,75	1,81	1,45									
45	Скв - 2	50,0	3	Глина черная твердая тонкослоистая	23,97					2,75	1,79	1,44					3,07	0,41	1,05		
50	Скв - 5	50,0	1	Глина твердая	16,64	48,45	25,71	22,74	0,40	2,73	1,94	1,66	20°30'	0,11							
50	Скв - 5	50,0	2	Глина твердая	34,54	49,12	26,81	22,31	0,35	2,73	1,97	1,46					2,47	0,43	0,57		
52	Скв - 5	70,0	1	Глина твердая	16,98	52,28	27,21	25,07	-0,41	2,73	1,80	1,54	23°	0,15							
53	Скв - 6	80,0	1	Глина твердая	18,48	53,13	28,10	25,03	-0,38	2,74	2,53	2,14					2,71	0,45	0,92		
54	Скв - 7	100,0	1	Глина твердая	19,54	52,38	27,15	25,23	-0,30	2,74	2,27	1,90	23°	0,13							
55	Скв - 8	120,0	1	Глина твердая	13,46	53,92	26,18	27,73	-0,46	2,74	2,06	1,82					2,57	0,31	0,44		

Table 3.2.

Results of determinations of physical and mechanical properties of rocky rocks of the Northern and Central Amantaytau deposits

Таблица 3.2

Результаты определений физико-механических свойств скальных горных пород месторождений Северный и Центральный Амантайтау

№ пробы	№ сква-жины, точки наблюде-ния	Глубина (интервал), горизонт отбора пробы, м	Кол-во образцов	Литотип	Плотность, г/см3	Предел прочности при сжатии, МПа: в крест слоистости	Предел прочности при сжатии, МПа: по слоистости	Предел прочности при растяжении, МПа	Модуль деформации	Коэффи-циент Пуассона	Модуль упругости	Угол внутрен-него трения	Сцепле-ние, МПа	Предел прочности при срезе со сжатием, $\tau_к$, МПа		
					ρ	$\sigma_{сж}$	$\sigma_{сж}$	$\sigma_р$	$E_д$	μ	E	$\varphi°c$	C	35°	35°	45°
1	2	3	4	5	6	7	8	9	10	11	12	13	14	15	16	17
3	т.н. 12	350	1	Сланец глинистый алевритовый	2,66	9,93 - 24,90 18,96										
3	т.н. 12	350	2	Сланец глинистый алевритовый	2,63							20°18′	0,008			
3	т.н. 12	350	3	Сланец глинистый алевритовый	2,64							19°18′	0,013			
4	т.н. 5	330	1	Сланец глинистый алевритовый	2,63							20°	0,011			
4	т.н. 5	330	2	Сланец глинистый алевритовый	2,65	35,87 - 50,32 42,50										
4	т.н. 5	330	3	Сланец глинистый алевритовый	2,63		8,74 - 27,73 18,24					19°18′	0,015			
4	т.н. 5	330	4	Сланец глинистый алевритовый	2,58	53,72 - 95,37 74,54			127	0,25	1,10					
4	т.н. 5	330	5	Сланец глинистый алевритовый	2,64		7,29 - 23,90 15,60		182	0,26	1,23					
5	т.н. 3	340	1	Сланец глинистый алевритовый	2,32	12,99 - 16,50 15,02										
5	т.н. 4	340	2	Сланец глинистый алевритовый	2,40		8,78 - 11,85 10,62									
7	т.н. 1	360	1	Сланец глинистый тонкосерицитовый	2,42							21°06′	7,22	8,80	11,01	11,74
7	т.н. 1	370	2	Сланец глинистый тонкосерицитовый	2,48							23°24′	4,19	5,25	8,50	7,39
7	т.н. 1	370	3	Сланец глинистый тонкосерицитовый	2,42							26°36′	5,40	7,06	10,09	10,86
7	т.н. 1	370	4	Сланец глинистый тонкосерицитовый	2,45				151	0,27	1,51					
7	т.н. 1	370	5	Сланец глинистый тонкосерицитовый	2,43			1,10 - 1,68 1,48								
8	т.н. 4	330	1	Сланец глинистый	2,36							19°48′	6,52	7,50	8,69	9,08
8	т.н. 4	330	2	Сланец глинистый	2,40							16°12′	2,21	2,56	2,92	3,12
8	т.н. 4	330	3	Сланец глинистый	2,41							17°24′	5,30	6,20	7,04	7,72
10	Скв - 8	26,0	1	Мергель	1,92		1,91 - 3,70 2,55		73	0,38	0,22					
11	Скв – 8	40,0	1	Мергель	1,97		3,21 - 6,33 4,76		28	0,32	0,36					
15	Скв - 8	82,0	1	Аргиллит	2,04		0,22 - 0,52 0,38		17	0,32	0,23	36°5′	0,26	0,40	0,65	1,01
16	Скв - 8	87,0	1	Мергель	2,10				25	0,41	0,27	28°	1,03	1,37	1,97	2,20
17	Скв - 8	100,0	1	Мергель	2,13							32°48′	0,74	1,05	1,22	2,06
18	Скв - 8	123,0	1	Сланец малопрочный	2,46		2,01 - 3,24 2,49		154	0,40	0,66					
19	Скв - 8	124,0	1	Сланец малопрочный	2,41		5,59 - 8,01 6,69		202	0,38	0,39					
20	Скв - 8	125,0	1	Сланец малопрочный	2,44		3,62 - 12,07 7,79		204	0,31	2,00					
21	Скв - 8	132,0	1	Сланец средней прочности	2,64		7,05 - 14,45 10,75					20°54′	9,73	11,84	13,13	15,75
22	т.н. 12а	345	1	Сланец средней прочности	2,60		9,70 - 17,11 13,11					19°12′	1,97	2,34	2,46	3,01
22	т.н. 12а	345	2	Сланец средней прочности	2,59							22°48′	3,34	4,15	5,18	5,75
22	т.н. 12а	345	3	Сланец средней прочности	2,61							19°36′	2,24	2,69	3,02	3,47
23	т.н. 19	320	1	Сланец средней прочности	2,52		26,26 - 47,41 34,35					28°24′	4,77	6,38	9,29	10,40
23	т.н. 19	320	2	Сланец средней прочности	2,56							20°36′	6,90	8,37	9,06	11,07
23	т.н. 19	320	3	Сланец средней прочности	2,54							20°30′	4,92	5,97	6,45	7,88

1	2	3	4	5	6	7	8	9	10	11	12	13	14	15	16	17
28	Скв - 6	43,0	1	Песчаник	2,15							33°24′	1,71	2,47	4,06	5,02
28	Скв - 6	43,0	2	Песчаник	2,13				136							
30	Скв - 6	62,0	1	Песчаник	2,36		4,11 - 8,59 5,69		338	0,30	1,52					
31	Скв - 6	81,0	1	Аргиллит	2,19		0,74 - 2,07 1,28		-	-	-					
32	Скв - 6	92,0	1	Аргиллит	2,18							34°	0,26	0,38	0,64	0,80
33	Скв - 6	100,0	1	Аргиллит	2,17		0,22 - 2,21 1,51					29°12′	0,32	0,44	0,64	0,73
34	Скв - 6	110,0	1	Аргиллит	2,17		0,22 - 0,50 0,35									
35	Скв - 6	120,0	1	Сланец	2,29		9,59 - 12,32 10,95									
37	Скв - 3	50,0	1	Мергель	1,89		1,82 - 3,56 2,51									
38	Скв - 3	51,0	1	Мергель	1,91		2,51 - 5,24		89	0,32	1,62					
39	Скв - 3	60,5	1	Мергель	2,04			1,02 - 1,28 1,15	97	0,31	1,48					
42	Скв - 3	110,0	1	Алевролит	2,25		0,51 - 1,00 0,72									
44	Скв - 2	40,0	1	Мергель	1,80				134	0,29	0,85					
44	Скв - 2	40,5	2	Мергель	1,81			0,20 - 0,84 0,52				25°30′	0,93	1,20	1,40	1,79
49	Скв - 2	110,0	1	Аргиллит слабо рассланцованный, прочный	2,23			0,11 - 0,93 0,27								
49	Скв - 2	110,0	2	Аргиллит слабо рассланцованный, прочный	2,30		0,47 - 0,55 0,52					19°12′	0,98	1,17	1,25	1,51
49	Скв - 2	110,0	3	Аргиллит слабо рассланцованный, прочный	2,27		0,22 - 1,13 0,75									
51	Скв - 5	50,0	1	Мергель темно-коричневый	1,95			0,18 - 0,36 0,27				21°24′	1,04	1,27	1,48	1,71
51	Скв - 5	50,0	2	Мергель темно-коричневый	1,91		2,66 - 5,08 3,70									
56	Скв - 5	140,0		Песчаник	2,56		8,14 - 8,54 8,34									

Table 3.3.

Results of determinations of physical and mechanical properties of loose rocks from the Northern and Central Amantaytau deposits

№ пробы	№ скважины, точки наблюдения	Глубина (интервал), горизонт отбора пробы, м	Кол-во образцов	Литотип	Влажность, %	Плотность, г/см³	Угол внутреннего трения грунта, град	Удельное сцепление грунта, МПа	Угол естественного откоса, град	Предел прочности при сжатии, МПа	Модуль деформации, МПа	Коэффициент Пуассона	Модуль упругости, кг/см²х10³	Коэффициент фильтрации, м/сут	Результаты гранулометрического состава горных пород							
															>10	10-5	5-2	2-1	1-0,5	0,5-0,25	0,25-0,1	0,1-0,05
					W	ρ	φ	C	α	σсж	$E_д$	μ	E	$K_ф$								
1	2	3	4	5	6	7	8	9	10	11	12	13	14	16	17	18	19	20	21	22	23	24
6	т.н. 23	370	1	Песчаник тонкозернистый	0,18	1,63	26°36′	0,003	31°								3,14	1,11	2,03	9,48	63,90	20,34
6	т.н. 23	370	2	Песчаник тонкозернистый	0,20	1,59			32°							12,51	10,58	3,79	3,72	7,20	50,03	12,18
6	т.н. 23	370	3	Песчаник тонкозернистый	0,21	1,62			30°													
12	Скв - 8	60,0	1	Песок мелкий	11,45	1,82	36°34′	0,006	38°						7,31	3,49	0,89	0,89	1,77	21,30	51,88	12,47
12	Скв - 8	61,0	2	Песок мелкий	11,97	1,79	36°36′	0,003	36°							1,50	0,07	0,27	2,55	26,79	48,56	20,27
13	Скв - 8	63,0	1	Песок	19,57	1,70	36°54′	0,005	33°						0,68	0,51	0,10	0,08	2,37	30,98	54,26	11,03
14	Скв - 8	80,0	1	Песчаник	4,76	2,48			34°	3,78 - 12,76 / 8,17	131	0,35	1,43			11,55	28,96	14,19	6,78	8,29	10,90	19,34
26	Скв - 6	26,0	1	Щебенистый грунт	7,47	2,09				0,17 - 2,22 / 1,12					4,68	20,83	23,03	13,66	7,33	10,64	10,86	8,96
27	Скв - 6	27,0	1	Песок мелкий	15,47	1,80	33°06′	0,005	28°							0,04	0,47	1,18	10,91	57,32	15,69	14,39
29	Скв - 6	50,0	1	Песок мелкий	13,50	1,89	35°64′	0,008	34°													
40	Скв - 3	80,0	1	Песчаник тонкозернистый (песок пылеватый плотный) светло-желтого цвета	14,68	2,24	27°30′	0,001			134	0,38	0,29	0,006	3,46	31,34	17,81	7,18	3,41	4,49	20,00	12,31
40	Скв - 3	81,0	2		16,77	2,13	25°12′в	0,001в						0,008	4,03	28,84	22,11	8,83	3,66	5,37	13,13	14,04
40	Скв - 3	82,5	3		16,77	2,11								0,058		22,51	10,68	3,99	4,72	7,84	39,52	10,74
46	Скв - 2	60,0	1	Песок пылеватый плотный влажный	12,19	1,62	35° в	0,010 в			9,86	0,37	0,83	0,625			0,02	0,07	1,67	32,35	54,03	11,86
47	Скв - 2	70,0	1	Песчаник тонкозернистый	13,77	1,88	31°	0,008			8,15	0,38	0,91			1,66	1,94	2,19	1,99	10,96	62,03	19,23
47	Скв - 2	70,0	2	Песчаник тонкозернистый	12,47	1,91				0,42 - 0,50 / 0,45 в												
48	Скв - 2	83	1	Песчаник тонкозернистый, глинистый	23,78	1,98	24°18′в	0,007в			9,21	0,36	0,87			13,67	22,66	9,25	5,29	5,29	35,34	8,61

Примечание: 35° в – испытания проб в водонасыщенном состоянии.

Note: 350 c - tests of samples in water-saturated condition.

Conclusions

1. The methodology of selection and preparation of monolithic rock samples for laboratory tests of physical and mechanical properties of rocks is substantiated. The strength characteristics of rocks are determined in the presence of widely developed tectonic disturbances, fracturing, layering of rocks, the degree of heterogeneity, the presence of cracks at the contacts, taking into account weakening factors.
2. The methodology of research of physical and mechanical properties of rocks is proposed. The obtained values of rock strength properties σ_{cj}, σ_p, cohesion and angle of internal friction allow to justify stable parameters of the quarry in conditions of quarry development, both the height of the ledge and the side, as well as the angle of their slope.
3. The limits of tensile strength, deformation properties, adhesion and angle of internal friction, Young's modulus of elasticity and Poisson's coefficient of rocks are determined on the basis of laboratory studies, which open the possibility of justifying the limiting parameters of the slope angle of the sides and ledges of the North and Central Amantaytau quarries.

CHAPTER 4

THEORETICAL SUBSTANTIATION OF PHYSICAL AND MECHANICAL PARAMETERS OF THE ROCK MASS ESTABLISHMENT OF LIMITING PARAMETERS OF THE SLOPE ANGLE OF SIDINGS AND BLASTING TECHNOLOGY

§ 4.1 Determination of physical and mechanical parameters of the massif according to the Hook and VNIMI methods

In order to carry out stability calculations, it is necessary, first of all, to establish parameters characterising physical and mechanical properties of the rock mass that composes the quarry sides. In this paper, two approaches are used for stability calculations and, accordingly, for determining the physical and mechanical characteristics of the rock mass.

The first approach is to apply the nonlinear Hooke-Brown strength criterion and the geological strength index (GSI). We used this strength criterion for stability calculations using the finite element method.

The second approach is to use the Coulomb strength criterion and to establish its parameters for the massif: cohesion and angle of internal friction in the massif. Using the Coulomb strength criterion, stability calculations were performed using the VNIMI limit equilibrium methods [82-83].

In both approaches, the parameters appearing in the strength criteria are determined depending on the characteristics of the massif disturbance studied as part of detailed exploration [60] and previous phases of this work [84-85].

When the parameters are established by Hooke's method, the Hooke-Brown strength criterion has the following form [86]:

$$\sigma_1 = \sigma_3 + \sigma_{ci}\left(m_b \frac{\sigma_3}{\sigma_{ci}} + s\right)^a \qquad (4.1)$$

where σ_a is the unconfined compressive strength, y, s and a are rock mass characteristics, which are found according to the following relationships:

$$m_b = m_i \exp[(GSI - 100)/(28 - 14D)] \qquad (4.2)$$

$$s = \exp[(GSI - 100)/(9 - 3D)] \qquad (4.3)$$

$$a = \tfrac{1}{2} + \tfrac{1}{6}\left(e^{-GSI/15} - e^{-20/3}\right) \qquad (4.4)$$

In the last three dependences *GSI* is the geological strength index proper; *mi* is the index characterising the increase of strength under lateral compression;

D is the index of disturbance of the massif by blasting.

The values of these characteristics adopted for the Amantaitau deposit are presented in Table 4.1.

Table 4.1 **Characteristics for rocks of the Amantaytau deposit**

	Sandstone, shale and siltstone interbeds	Cataclasites	Faults
GSI	25	22	16
mi	14,3	14,3	14,3
D	0-1	0-1	0-1

The GSI value was taken as the average of the three determinations made in Chapter 2 of the paper.

The value of *mi is* taken as a weighted average of the tabulated values (19 for sandstones, 9 for siltstones, 6 for shales) recommended by Hook [87] for the interbedded rocks of the field. The weights are assumed to correlate with the total interval lengths for the wells in the database. The total length of sandstone intervals in the boreholes is 48376.2 m, siltstone - 27180.8 m, shale - 9805.5 m.

The index of disturbance of the instrument massif by blasting operations is taken as D = 0 in the depth of the massif and D = 1 at the surface of the quarry face. The width of the zone disturbed by blasting with D = 1 is taken equal to T = 1.5 ΄H = 23 m, where H is the height of the working ledge, according to the recommendations of Hook [88].

The free compression strength is assumed to be equal to the uniaxial compressive strength due to the lack of results of tests for triaxial compression. The average uniaxial compressive strength of rocks of the Paleozoic complex is 50.5 MPa [85].

Rock parameters not relevant to the Hooke-Brown strength criterion are discussed below.

According to the VNIMI methodology, stability calculations are performed using Coulomb's strength criterion and this requires the determination of parameters:

- cohesion of rocks in the massif (C);
- angle of internal friction in the array (φ);
- volumetric weight of rocks (γ).

The values obtained for the sample are taken as the angle of internal friction and volumetric weight of rocks in the massif.

The cohesion in the rock and semi-rock massif is determined using the known empirical dependence of VNIMI [82]:

$$C_M = \frac{C_0 - C'}{1 + a' \ln\left(\frac{H}{l_T}\right)} + C' \qquad (4.5)$$

In this dependence H is the highest height of the projected quarry face, in our case equal to 410 m.

Where: a' - tabular coefficient depending on the strength of rocks in a piece and on the nature of their fracturing. In the conditions under consideration, according to [83, 89], taking into account the mechanical properties of rocks and the characteristic of their fracturing, this coefficient is assumed to be $a' = 3$ for rocky rocks; lt - size of the structural block (distance between cracks). According to the conducted studies [85] the average distance between cracks in the Paleozoic rocks of the deposit is 6.48 cm; c_0 - cohesion in a piece. The average value of cohesion in a piece for sandstone, shale and siltstone interlayers is 622 t/m^2 [85]; C - cohesion along the fracture, taken for rock rocks as 1.2 t/m^2.

Table 4.2 shows the calculated characteristics of rocks in the massif found using this dependence.

Table 4.2 Calculated values of physical and mechanical properties of the massif

Breeds	Clutch, t/m^2	Angle of internal friction,	Volume weight, t/m^3
Sandstone, shale and siltstone interbeds	23,8	21,5	2,52
Fault rocks	6,0	21,5	2,52
Cataclasites	6,0	21,5	2,52
Dealluvial clay	3,77	19,2	2,04
Alluvial clay	6,42	21,6	2,02
Sand	0,25	33,0	1,80
Mergel	47,9	27,0	1,95

Separate tests and surveys of fault rocks and cataclasites have not been carried out, so the cohesion on them is taken from the data of analogue deposits [61], and the internal friction angle and volume weight are taken from the data of rock tests [85].

Rock properties of loose sediments are taken from laboratory test data [85]. A scaling factor (similar to the structural weakening coefficient for rock) λ =

0.5 is introduced into the rock cohesion.

The friable rock parameters specified in Table 2.2 were also used in the numerical modelling using the finite element method.

The components of crack strength in the calculations, based on the results of laboratory tests, are taken as follows: cohesion C' = 1.2 t/m^2, angle of internal friction φ = 19.60.

To compare the obtained strength curves for rocks of the Palaeozoic complex of the deposit, their graphs were plotted (Fig.4.1). Normal stresses are plotted on the horizontal axis, and tangential stresses are plotted on the vertical axis.

Comparison of the plots in Fig. 4.1 shows that, in general, the Coulomb strength envelope, constructed partially from the characteristics obtained by the VNIMI method, lies within the range of possible values of the Hooke-Brown strength. The linear envelope coincides to the greatest extent with the Hooke-Brown strength curve at low values of normal and tangential stresses and at the explosion disturbance factor D=1. On the one hand it means that the slope stability margin calculated by two different methods will give close to each other results. On the other hand, the stability of the quarry face should not be very different in both cases, because the value of *D* decreases in the direction to the depth of the massif, where the stresses will be the highest.

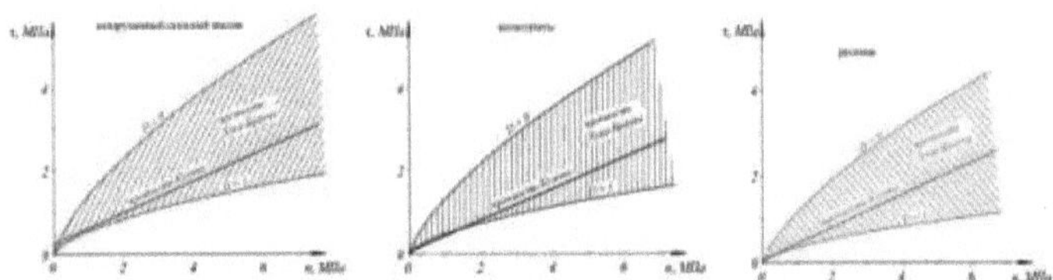

Fig. 4.1. Comparison of Coulomb and Hooke strength curves of rocks from the Northern and Central Amantaytau deposits

The indirect comparison also confirms that the selection of strength properties of rocks in the massif is correct.

§ 4.2 Estimation of the stability of the quarry sides in the design finite element method and limit equilibrium method equilibrium

In international practice, it is recommended that limit equilibrium calculations be verified by numerical modelling [89]. The need for numerical modelling is due to the impossibility of taking into account the complex stress state under the assumption of absolute rock stiffness using limit equilibrium methods.

Stability calculations were carried out using the finite element method. Midas GTS NX software environment was used for modelling. Appendix 4 contains

a certificate of compliance of this programme with regulatory requirements.
The solution is performed in two-dimensional formulation. The model of rock behaviour is elastic-plastic with yield strength corresponding to the Hooke-Brown strength curves for rocks of the Paleozoic complex and Coulomb for rocks of the upper friable strata.
Calculation of the stability factor of the quarry face in the design contour is made by the strength reduction method (SRM). This method consists in step-by-step reduction of strength characteristics of materials in the model and as some elements are no longer provided with Coulomb strength condition, they are excluded from the calculation. The coefficient by which the characteristics are reduced, at which the calculation ceases to converge (i.e. the model fails), is taken as the stability margin.
The dimensions of the model were chosen to ensure that the stress state around the excavation is not influenced by boundary conditions.
The calculations take into account the effect of the natural stress field studied by in this paper, which is given in the following form:

$$\sigma_{гор.мерид.} = 3МПа + 1{,}14\gamma H, \qquad (4.6)$$

$$\sigma'_{гор.мерид.} = 2МПа + \gamma H. \qquad (4.7)$$

In the model, this dependence is approximated when given through the lateral pressure coefficients κ_0 and given in the form of reduced dependences when using the initial equilibrium force (EEF) function.
Fig. 4.2 shows as an example the appearance of a Midas GTS NX section adjusted in the preprocessor (i.e. before the calculations are started). In the same way, 12 profile line transects, shown in Fig. 4.3, were tuned. 4.3.

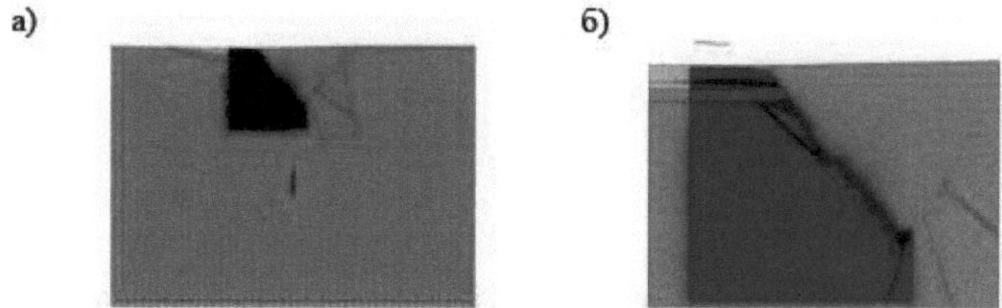

Fig. 4.2. Model prepared for the start of calculations: a) general view of the model; b) detail of the calculated quarry face

At the boundaries of the model, the displacements are limited: *x-axis*, y-axis along the lower horizontal boundary, x-axis along the lateral vertical boundaries.
The central part of the model, visually highlighted in Fig. 4.2, is divided into a finer grid. The elastic-plastic behaviour of the rocks in this part of the

model is chosen, while in the rest of the model the rocks are elastic.

The modulus of elasticity and Poisson's ratio assumed in the model are given in Table 4.3. The modulus of elasticity of loose and semi-friable rocks is assumed to be equal to the same value in the piece. For rocky rocks, a recalculation to the massif was performed in accordance with the empirical relationship proposed by Hook and Diedrichs [86]:

$$E_{rm} = E_i \left\{ 0{,}02 + \frac{1 - D/2}{1 + \exp[(60 + 15D - GSI)/11]} \right\} \qquad (4.8)$$

where E_i is the modulus of elasticity of the rock in the piece.

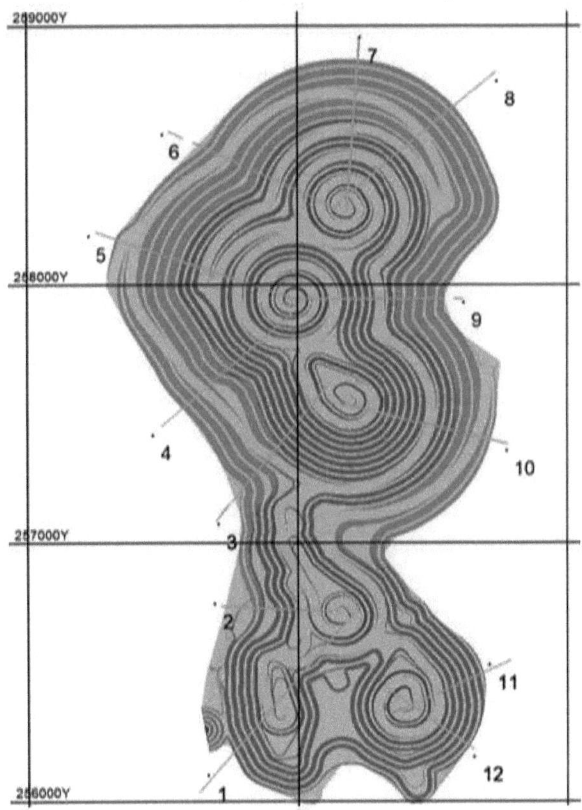

Fig. 4.3. Quarry plan with the location of the design cuts sections

Table 4.3.

Mechanical characteristics of the model

Breeds	Modulus of elasticity, MPa	Poisson's ratio

Sandstone, shale and siltstone interbeds	997,8 (421,0)	0,31
Fault rocks	633,3 (372,3)	0,31
Cataclasites	844,1 (400,2)	0,31
Dealluvial clay	2,7	0,47
Alluvial clay	2,7	0,41
Sand	9,1	0,37
Mergel	74,3	0,34

Using the parameters described above, modelling of the stress-strain state of the massif by the finite element method was carried out when setting the quarry sides to the design contour. Along the slope of the face, a zone with reduced characteristics with the disturbance index D=1 is selected. The rest of the model is defined with zero disturbance index.

Modelling was carried out for 12 sections, with stability margin calculation using the strength reduction method.

The calculation was performed in stages. The first stage involved modelling of the stress-strain state of the massif prior to quarrying, and the second stage involved calculation during rock excavation in the design contour with strength reduction.

The modelling showed that none of the sections has a stability margin greater than unity. The features of Midas GTS NX software limit the possibility of calculating the stability margin with the reserve coefficients less than unity when specifying lateral horizontal (tectonic) stresses; therefore, to display the stress-strain state, additional modelling of each section was performed with the tectonic stresses specified through the lateral pressure coefficients κ_0.

Graphical results of the performed numerical modelling for all transects can be found in Appendix 5.

Calculation of stability according to the method of VNIMI for the conditions of Amantaytau deposit is performed by the method of vector addition of forces. This method is based on the theory of limit equilibrium of the massif and satisfies the conditions of equality of both the sum of moments and the sum of force vectors vertically and horizontally. Also this method allows to accept non-vertical boundaries of blocks and thus to take into account the secondary most stressed surfaces and weakening surfaces falling deep into the massif.

The calculations were performed according to the special schemes recommended by the "Rules for ensuring stability...", taking into account the structural and geological peculiarities of the quarry sides. For the quarry faces located in the hanging side of the deposit, the interaction between the blocks was calculated taking into account the possible location of extended shale fractures between the blocks. The stability reserve of the quarry faces located in the lying side of the deposit is calculated taking into account the possible location of a crack along the potential sliding surface.

All calculations were performed graph-analytically. Appendix 6 contains calculation sections with vector diagrams of acting forces, diagrams of division into blocks of the collapse body, tables with moduli of block vectors and other explanations of the calculation.

The vector diagrams were constructed at two different stability margin factors, which are analogues of the strength reduction factor in FEM modelling, as they are introduced in the strength characteristics. Next, the non-convexity of the vector diagram (deficit or surplus of restraining forces) is calculated. After that, the stability margin factor is determined graphically by linear interpolation, at which the zero looseness will be achieved.

The graphs (Fig.4.4) show the obtained stability reserve coefficients of the carriage sides in the design contour for 12 transects.

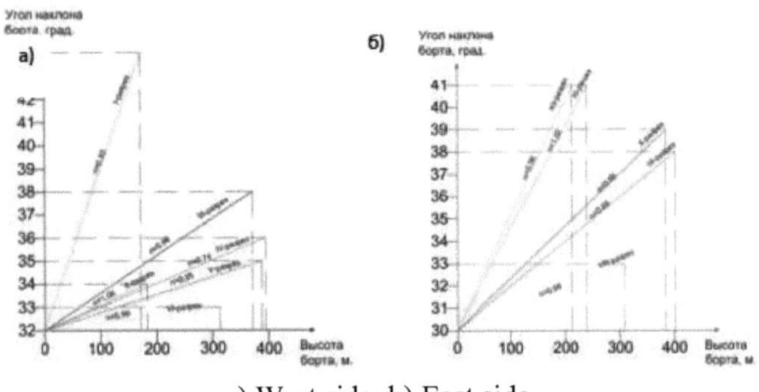

a) West side; b) East side.

Fig.4.4. Graph of calculated stability reserve coefficients of the Amantaytau quarry face in the design contour for 12 sections.

The results of the calculations show that the stability margin of the sides is not sufficient. That is, the current design configuration of the quarry should be significantly revised. The results of the calculations generally coincide with the results of finite element modelling.

§ 4.3 Determination of parameters of stable ledges

The primary structural sections of the Amantaytau deposit were identified. A structural section is a limited area of the massif that has a structural structure distinctive from neighbouring sections.

Before zoning of the quarry sides by the recommended slope angles, it is necessary to consolidate the primary structural areas and allocate the design sectors of the quarry sides areas, distinguished by the mutual orientation of the side and fracture systems.

The selection of design sectors can be done based on the rules of kinematic analysis. The stable slope angles of the scarp slopes are then determined using a probabilistic approach, kinematic analysis or simplified limit equilibrium calculations.

In this paper, the calculation of stable ledges is performed using kinematic analysis and limit equilibrium methods, since the use of probabilistic calculations is not possible due to the lack of statistical data on the fracture survey on the quarry face.

It should be added that the calculation of stable ledges is made for the conditions of application of special technology of drilling and blasting operations in the contour zones. That is, without the introduction and development of special technology, the given parameters of ledges and sides of the quarry are not achievable. According to our estimations, without application of special blasting technology, the ledges will be flattened to 40-450 at the height of 30 m ledge. This will entail the reduction and failure of safety berms and subsequent high frequency of scree, rockfalls and localised cave-ins dangerous for personnel and equipment working at the quarry.

Appendix 7 shows the quarry plan with design sectors with signed recommended (stable) parameters of the quarry ledges in the rock strata. The sectors are coloured according to the most likely scenario of loss of stability of the ledges. The colour legend is also given in Appendix 7.

The maximum stable angle of the ledge in the absence or favourable location of weakening surfaces was calculated by the limit equilibrium method with the adoption of the Hooke-Brown criterion at the blasting disturbance coefficient D=0.7 and was 650. Sufficient stability reserve coefficient at calculation by the given method is accepted equal to p_n = 1,3. The verification calculation with Coulomb characteristics showed stability with a safety factor equal to n=2.1. This difference is explained by the large difference in the maximum allowable tangential stresses in the curves at low normal stresses and, accordingly, at low slopes.

Wedge-shaped faults were also analysed using the limit equilibrium method in Geomix software. Figure 4.5

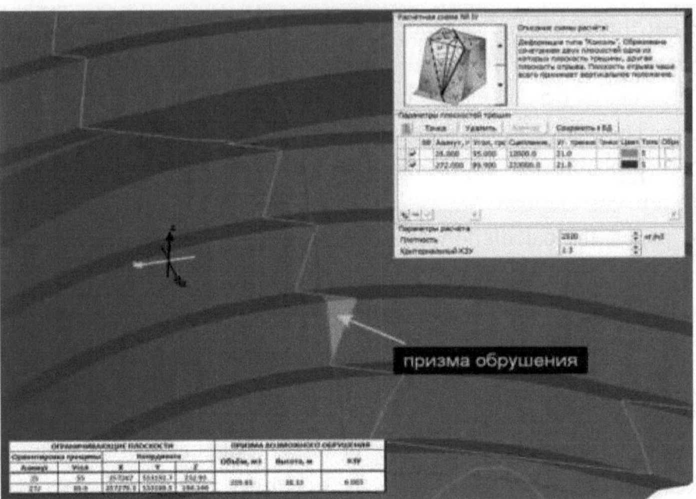

Fig.4.3 Example of calculation of stability of wedge-shaped disturbances

Since the small-scale disturbance of the deposit's rocks is represented only by shale, i.e., in fact, by a single fracture system, wedge-shaped collapses at the open pit will occur with detachment on one of the sides. Based on the results of the analysis, it is concluded that it is not necessary to allocate design sectors for wedge-shaped cave-ins. The dip angles of shale formation in the deposit are mainly such that the volume of possible collapses will be small. If the slope angle changes smoothly, berms will be preserved at the boundaries between the sectors where wedge-shaped collapses are possible.

For slopes dangerous by plane collapse, slippage is assumed by layering when the friction angle of the crack is exceeded. The accepted parameters are checked by limit equilibrium calculations according to special schemes 2a and 2b of the II group of typical calculation schemes of the "Rules for ensuring stability...". If the verification calculations reveal insufficient reserve, the slope angle is assumed to be slightly lower.

For slopes disturbed by reverse fall cracks, located mainly on the eastern side of the quarry, a correction to the angle of stable slope calculated according to the VNIMI method [83] was used. The ledges are considered to be tipping hazardous when the fracture dip angle is greater than 600.

As for loose strata, the parameters of ledges to be built in loose strata can be calculated by the method of limit equilibrium with the selection of an angle that ensures stability. Angles of ledges in sand should preferably be taken according to the angle of internal friction. Recommended parameters of ledges by rocks are shown in the graph (Fig. 4.6).

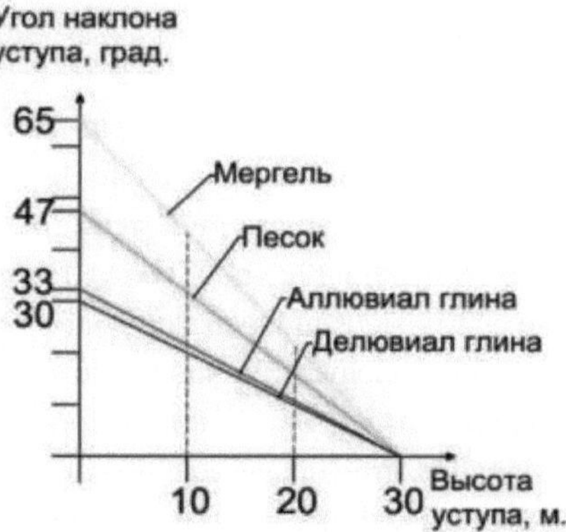

Fig.4.6: Graph of dependence of ledge height stability on ledge slope angle

Quarry zoning in the project contour with indication of the parameters of ledges composed of loose rocks is given on the plan of the quarry in the project contour in Appendix 7.

§ 4.4 Finding general angles of stable pit sides

On the basis of the obtained data on physical and mechanical properties of rock massif and structural and geological conditions, stability calculations for selection of general angles of the pit sides have been performed. On the basis of the calculations the angles of inclination of the sides were determined, providing normative stability margin of the side n=1,3. The calculations were performed on the sections built on the profile lines shown in Fig. 4.3. 4.3. Zones with similar characteristics around the profile lines are limited by similarity of mining and geological conditions, such as the height of the side, the angle of inclination of the side, and the composition of the rocks composing the side.

The results of calculations in the form of zoning of quarry sides by stability indicators are given in Appendix 8.

The possible influence of tectonic disturbances on the stability of the pit walls was also analysed in the development of the recommendations. In general, based on the available geological and structural model, no pronounced unstable structures have been identified, but attention should be paid to two areas that may be unstable. These are the areas on the south-western side of the south-eastern excavation and on the eastern side of the northern deepest excavation of the open pit. These hazardous areas should be safeguarded in the new pit design where possible, and further investigation and assessment of the stability of these areas as they are exposed by the pit excavation should be undertaken.

Conclusions

1. The calculations with application of nonlinear Hooke-Brown strength criterion and geological strength index (GSI) by finite element method, also with application of Coulomb strength criterion by limit equilibrium methods obtained strength curves by different methods showed their similarity.
2. The stability of the quarry sides in the design contour was assessed in two ways. Modelling by finite element method and calculations by limit equilibrium method (method of vector addition of forces). The results of the stability assessment signalled insufficient stability of the quarry sides in the design contour. The obtained stability reserve coefficients lie in the range of 0.66-1.06.
3. The general inclination angles of the pit sides have been determined, which will ensure their long-term stability.

CHAPTER 5

DEVELOPMENT OF BLASTING TECHNOLOGY OF NEAR-CONTOUR RIBBONS IN THE CONTOUR ZONE OF THE SIDEWALL OF THE QUARRY AND SLOPE DESIGN IN THE ULTIMATE LOCATION

§ 5.1. Development of the methodology of calculation of blasting by the method of near-contour belts

Conducted studies of physical and mechanical properties and deformations on the contour of the quarry board at zoning of the quarry field on stability of slopes at approach to the limit contour it is necessary to change the technology of drilling and blasting operations.

Despite the ever-increasing scale of blasting and its impact on slope stability, most open pit mines have not yet developed, and therefore have not applied, effective measures to reduce the seismic effect when approaching the limit contour.

One of the criteria for determining the amount of simultaneously detonated explosives is the value of the seismic hazard measure, at which the residual deformations of the rocks composing the ledges and sides of the quarry are practically excluded.

In the work of V.N. Popov and B.N. Baikov [90] it is established that the vertical displacement (mm) can be determined by the formulas:

for soft kaolinised rocks

$$\Delta h = 55/L - 1{,}25 \qquad (5.1)$$

rocky

$$\Delta h = 106{,}6/L - 1{,}33 \qquad (5.2)$$

where L is the distance to the blast hole, m.

For horizontal displacements (mm) is determined by the following equations:

for soft kaolinised rocks

$$\Delta l = 88/L - 2 \qquad (5.3)$$

rocky

$$\Delta l = 1187{,}5/L - 2{,}5 \qquad (5.4)$$

Since the limiting error of measuring vertical Δh and horizontal Δ l deformations with the adopted measurement technique is 3 mm, the maximum permissible value of the measure of seismic hazard for all cases will be determined from the equation

$$\Delta l, \Delta h = \exp(a \cdot \rho - b) \quad (5.6)$$

where *a, b* - coefficients taking into account the type of deformation and location of the measurement point relative to the explosion Table 5.1; ρ - measure of seismic hazard, kg$^{1/3}$/m.

Table 5.1.

Values of coefficients *a* and *b*

Point location	Type of deformation			
	Horizontal displacements		Vertical displacements	
	a	*b*	*a*	*b*
On the horizon of the blasted ledge	9,5-13,5	1,8-9,7	9,5-12	1,8-8,7
One horizon above the ledge to be blasted	12,9	9,8	5,9	7,3
Two horizons above the blasted ledge	22	13,1	13,9	9,5

After substituting Δh and Δl equal to 0.003 m into equations (5.6). Taking into account the obtained values of ρ, the safe distance for contour ledges from the explosion site to the protected object should be determined by the formula

$$r_o = K_c \sqrt[3]{Q} \quad (5.7)$$

where *Q is the* mass of simultaneously exploded explosives, kg; K_c is a value inversely proportional to the seismic hazard, i.e., *Ks.*

$$K_c = \frac{1}{\rho} \quad (5.8)$$

The numerical values of this coefficient depend on the criterion for the type of displacement, location of the protected object, type of rocks and nature of fracturing (Table 5.2).

Table 5.2 K_c values for different conditions

Location of the protected object	Average size of elementary block rib, m	*Ks*	ρ
On the horizon of the blasted ledge	Up to 0.1	8,7	0,115
	0,1-0,3	6,2	0,161
	0,3-0,6	3,76	0,266
	0,6-2	3,02	0,331
	2	2,8	0,357
On the horizon above the blasted ledge	Up to 0.1	8,22	0,122
	0,1-0,3	5,87	0,170
	0,3-0,6	3,56	0,281
	0,6-2	2,85	0,351

	2	2,65	0,377
Two horizons above the ledge to be blasted	Up to 0.1	7,89	0,127
	0,1-0,3	5,61	0,178
	0,3-0,6	3,42	0,292
	0,6-2	2,74	0,365
	2	2,54	0,394

Due to the fact that the rocks at the horizon of the blasted ledge are subjected to the greatest deformation, *the* calculation of the amount of simultaneously detonated explosives should be carried out taking into account the coefficient K_c corresponding to these conditions. The closeness of the coefficient values for vertical and horizontal displacements allows to use average values with an error of 7-9%.

Then the amount of simultaneously detonated explosive at the approach of blasting to the limit contour

$$Q=\left(\frac{r_o}{K_c}\right)^3 \qquad (5.9)$$

where K_c *is the* average value of the coefficient.

According to the results of studying the effect of mass explosions on the deformation of contour ledges, the method of Y.I. Turintsev determines the value of massif weakening depending on the number of simultaneously detonated explosives Q and the distance to the place of explosion L_B:

$$\tau = 0{,}45\cdot 10^3 Q/L_B^2 \qquad (5.10)$$

Since on the limit contour τ = 1, for these conditions the dependence (5.10) will take the form

$$Q = L_B^2/4{,}5 \qquad (5.11)$$

where L_B is measured in tens of metres.

Thus, the data of Table 5.4 allow us to determine the amount of simultaneously detonated explosives depending on the structure of the massif and the scheme of construction of non-working ledges in the limiting contour of the quarry face.

Proceeding from the provision of the minimum zone of intensive deformation, it is possible to determine, according to the revealed regularities and the data of Table 4.8, the optimum specific consumption and quantity of explosives per 1 m of the work front on the basis of dependences

$$L = Z(q_{\phi} - q_o)^{\lambda} \rightarrow \min \qquad (5.12)$$

$$L = \beta(q_{м} - q_{м.o})^{\varepsilon} \rightarrow \min \qquad (5.13)$$

where, L - the length of the zone of intensive deformation in the direction in the cross direction of the scarp extension, m; qf, $q^{(}_{o)}$ - respectively actual and optimal specific expenditure of explosives, kg/m^3; $q_{(m)}$, $q_{m\cdot o}$ - respectively actual and optimal amount of explosives, falling on 1 m of the front of works, kg/m; Z,, β, ε - coefficients, taking into account the properties of rocks and fracture tectonics of the massif (Table 5.3).

Table 5.3.

Values of the coefficients included in equations (5.12) and (5.13)

Careers	Rock strength	q_o	$q_{m.o}$	Z	λ	β	ε
Quarries Central and	9-17	0,4	108	30,29	0,625	3,269	0,292
Northern fields	8-14	0,37	58	10	0,333	2,78	0,213
Amantaytau	1-6	0,3	25	16	1	0,15	1

According to the established optimum consumption of explosives, which allows to exclude legitimate deformation of the massif, it is possible to determine the width of the contour zone R, which is the distance from the upper edge of the worked out ledge to the points towards the stationary side, where the displacement does not exceed 3 mm.

In general terms, the amount of simultaneously detonated explosives

$$Q = l_{л} \cdot h \cdot L_{л} \cdot q_o \qquad (5.14)$$

where $l_{(}l_{)}$- width of the strip, m; h - height of the ledge, m; L_n - length of the strip, m; q_o - optimum specific consumption of explosives, kg/m$^{(3)}$.

Substituting expression (5.14) into equation (5.6), we have

$$R = K_c \sqrt[3]{l_{л} \cdot h \cdot L_{л} \cdot q_o} \qquad (5.15)$$

At the same time, the amount of explosives consumed per 1 m of work front can be represented analytically as

$$q_{м} = l_{л} \cdot h \cdot q \qquad (5.16)$$

where q is the specific consumption of explosives.

Hence, taking into account the established optimal parameters q_o and q^, it is possible to determine the maximum width of the contour belt, providing minimum destructibility of the contour massif, according to the formula

$$l_{л} = q_{м.o} / (h \cdot q_o) \qquad (5.17)$$

Width of the developed strip depending on the bottom resistance line w, the number of well rows *n* and the distance between well rows B

$$l_s = w + (n-1)B \qquad (5.18)$$

On this basis, the width of the contour zone is determined by the transformed formula (5.15):

$$R = K_c\sqrt[3]{[w+(n-1)B]\cdot h\cdot L_s\cdot q_o} \qquad (5.19)$$

Taking into account the data of Table 5.5 of physical and mechanical characteristics of rocks and dependences (5.12) and (5.13), it is easy to obtain an equation of the form

$$R = A\sqrt[3]{L_s} \qquad (5.20)$$

where *A* - empirical coefficient, for the conditions of the Central quarry from 11.5, Northern up to 18.

Expression (5.20) allows at a given distance from the blast site to the limiting contour of the pit face to determine the dimensions of the blasted block along the front of works (Fig. 5.1).

From equation (5.18) it is possible to determine the number of rows of wells in the blasted block at different height of the ledge:

$$n = \frac{q - w\cdot h\cdot q_o}{h\cdot q_o\cdot B} + 1 \qquad (5.21)$$

The mass of charges in the boreholes is determined taking into account the fracturing of rocks according to the formulas (5.22), (5.23) and (5.24).

$$Q = \frac{\pi\cdot\Delta\cdot\rho\cdot c^2\cdot d\cdot a}{4P}\left[H\cdot\sigma_p + (3\sin\alpha + H\cos\alpha)\cdot(\sigma_c\sin\alpha + H\cdot\gamma\cdot\mu)\right] \qquad (5.22)$$

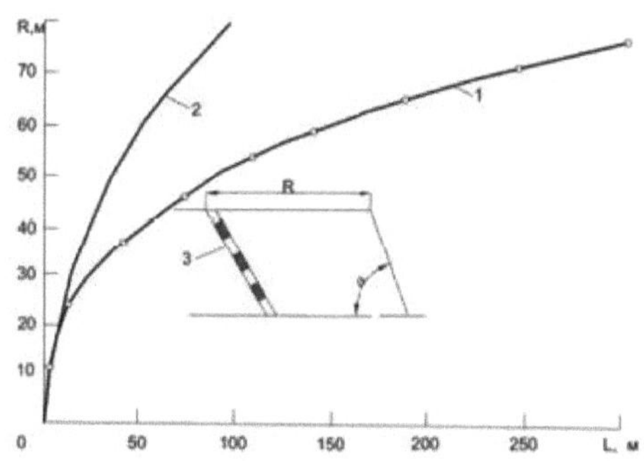

1 - Central pit; 2 - Northern pit; 3 - design contour; *a* - *angle* of inclination of the ledge.

angle of inclination of the ledge.

Fig. 5.1. Graph of change of the width of the contour zone *R* as a

function of the length of the blasted ledge along the front L depending on the length of the blasted ledge along the front *L*

A fractured massif should be considered as a sum of elementary monolithic blocks formed by different fracture systems. When exploding fractured rocks, the amount of explosives in the borehole will depend on the degree of fracturing. Therefore, the formula (*5.22*) for a fractured massif will have the form

$$Q_{mp} = Q / K_{mp} \tag{5.23}$$

where, K_{tr} - rock fracture coefficient

$$K_{mp} = 1 + l / l_{б} \tag{5.24}$$

The distance between wells in a row to ensure minimum destruction of the rock mass of the non-working ledge,

$$a = Q_T / (w \cdot h \cdot q_o) \tag{5.25}$$

where Q_T *is the* mass of the charge in the borehole, taking into account the natural fracturing of rocks.

Taking into account the values of *w* and q_0, we obtain

$$a = Q_T / K_u \tag{5.26}$$

where K_i - coefficient, for different conditions of quarries equal from 54 to 58.

The maximum specific charge in a group of charges is determined by the transformed equation:

$$\beta = Q / L = V^2 r^2 / (K_1^2 e^{-0,06r}) \tag{5.27}$$

Taking into account that the speed of permissible rock vibrations in the ledges is 24 cm/s, for the conditions of the quarry we have:

for rocks with $f = 9 \div 11$

$$\beta = 0{,}0050 r^2 / e^{-0,06r} \tag{5.28}$$

for rocks with $f = 11 \div 13$

$$\beta = 0{,}0059 r^2 / e^{-0,06r} \tag{5.29}$$

for rocks with $f = 13 \div 16$

$$\beta = 0{,}0073 r^2 / e^{-0,06r} \tag{5.30}$$

Hence, taking into account formulas (5.9), (5.29) and (5.30), by which specific charges and the number of simultaneously exploded explosives are determined, the maximum work front for different conditions is found from expression.

$L_з = Q / \beta$ (5.31)

The number of well rows in the contour strip can be adjusted by β.

Drilling and blasting parameters were determined according to the adapted methodology of B.R. Rakishev [91]. Depending on the viscosity and degree of fracture development, rocks are divided into three categories of explosivity (Table 5.4).

Table 5.4.

Classification of rocks by explosiveness

Explosivity category of rocks	Explosive	Moderately explosive	Hard-to-explode
Specific consumption BB, kg/m^3	0,36	0,4	0,44

The sole resistance line (l.s.p.p.) is determined by the formula

$$w = w_o + 0{,}5H$$ (5.32)

where w_O is the component of L.s.p.p., depending on the explosivity category of rocks and type of explosives; 0.5 is the coefficient that takes into account the increase in L.s.p.p. depending on the height of the ledge; *H* is the height of the ledge, m.

The distance between the wells is determined by the dependence

$$a = a_o + 0{,}25H$$ (5.33)

where aO - component of the distance between wells, depending on the category of rock explosivity and type of explosives; 0.25 - a coefficient that takes into account the increase in the distance between wells depending on the height of the ledge.

The perebore for all rock types is the same and is as follows

$$K = p \cdot H$$ (5.34)

where *p* is a coefficient depending on the explosivity category of rocks. The length of the borehole is determined by the dependence

$$l_{заб} = Z \cdot w$$ (5.35)

where *Z is the* bottom hole coefficient.

Estimated specific consumption of explosives is taken on the basis of the following equality

$$C_p = q_1 - q_2 H$$ (5.36)

where $q1$ - specific consumption of explosives necessary for normal crushing of the massif at the height of the ledge H = 6 m and *a*, *w*, *K*, calculated by the formulae (5.32) ^ (5.35); q_2 - a coefficient that takes into account the reduction of the specific consumption of explosives depending on the height of the ledge *H* (Table 5.5).

The mass of the borehole charge is determined by the formula

$$Q = C_p a \cdot w \cdot H \qquad (5.37)$$

Table 5.5.

Values of empirical coefficients for different rocks in terms of explosivity

Explosivity category of rocks	*Wo*	*ao*	*Z*	*q* i	*q2*	*u*	*p*
Explosive	5	7	0,7	0,31	0,00	1,15	0,10
Moderately explosive	4	5	0,75	0,41	0,01	1,08	0,15
Hard-to-explode	3	5	0,8	0,68	0,02	1	0,2

Blasting was carried out using the multi-row short-delay blasting method (Table 5.6).

The need to maintain the stability of the sides for a long time requires the application of effective methods of development of the contour belts by drilling and blasting method at the designed angles, taking into account the mining and geological peculiarities of each deposit.

Taking into account the conducted zoning of the Amantaytau quarry sides allows to determine the areas that have angles not corresponding to the physical and mechanical properties of rocks. In such areas it is recommended to carry out BWR by the method of multi-row short-delayed blasting with the use of slot-forming inclined contour boreholes with the use of trunnated hryland charges.

Table 5.6

BWR parameters for single-row and multi-row blasting and borehole diameter of 243 mm

borehole diameter equal to 243 mm

Explosivity category of rocks	Height of ledge, m	Parameters of the BVR		
		Distance, m		
		between rows between wells	Between wells in a row	Overbore, m
Explosive	10	7,5	7,5	1,5
	14	8,5	8,5	2'
	18	9	9	2
	22	9,5	9,5	2
Moderately explosive	10	6,5	6,5	1,5

	14	7	7	2,5
	18	8	8	3
	22	9	9	3
Hard-to-explode	10	5	5	2
	14	6	6	3
	18	6,5	6,5	3,5
	22	7,5	7,5	4

Parameters of drilling and blasting operations at development of the contour belt by inclined contour slot-forming wells are determined according to the methodology, according to which the line of the least resistance is defined as:

$$w_p = 0{,}87\sqrt{p/(m \cdot K)} \quad (5.38)$$

single charge mass

$$Q = C_p w_p a H_y / \sin\alpha \quad (5.39)$$

bottom hole length

$$l_{заб} = (20 \div 25) d_{зар} \quad (5.40)$$

where p - capacity of explosives in 1 m of well, kg; C_p - estimated specific consumption of explosives, kg/m^3, determined by Table 5.5; a - distance between wells in a row, m; N_u - height of the part of the scarp on which the wells were drilled, determined graphically by the section based on the line of least resistance and the angle of inclination of the well, m; α - estimated angle of slope of the scarp (angle at which the wells were drilled), degree; d_{3ap} - diameter of the charge.

§ 5.2. Development of technological schemes of blasting operations by the short-delayed blasting method on the contour of quarry sides.

Known studies have established that the use of inclined boreholes allows to make a qualitative design of ledges, put in a limiting position, as well as in those cases, when during the mining of ledges the width of the contour protection zone for one or another reason precludes the use of other methods of scarping. The method is recommended for use in areas with different strength and structure of the rock massif.

Development of the last contour protection zone is recommended by drilling and blasting three rows of vertical and one row of inclined boreholes at the index $n>1$. At the index $n=0,9$ *it* is recommended drilling and blasting of two rows of vertical and one row of shortened vertical and one row of inclined wells. At the index $n=0,8$ *it* is recommended to drill and blast two rows of vertical and two rows of inclined boreholes. To put the ledge in the limit position, inclined boreholes are used with a diameter of 243 mm, located on the design contour, at a design angle at a distance of 3-3.5 m from each other. The auxiliary row of inclined boreholes is located at a distance of 5 m from the design top edge of the ledge.

The adopted order of blasting ensures the reduction of seismic effect of the mass explosion on the massif. The grid of blast holes is 6x6 m for hard-blasted rocks and 7x7 for medium-blasted rocks, thus achieving the practical absence of spalling beyond the design contour of the ledge.

Blasting patterns and deceleration time have a significant impact on the results of scarping. In the conditions of Amantaytau open pit three schemes of scarping were tested:

- with three rows of vertical boreholes and along the design contour with single-row deviated boreholes to create a set scarp slope;
- with two rows of vertical boreholes, one row of shortened vertical boreholes and along the project contour single row slanting boreholes using short auxiliary and vertical jacking boreholes to better crush the part of rock located between the slanting boreholes and vertical jacking boreholes;
- with two rows of vertical wells and two rows of deviated wells, one row of which was drilled along the project contour line.

When applying the proposed blasting schemes, short-delayed blasting is used with a delay between rows equal to 35-42 ms. In the first case, in the beginning, all vertical boreholes are blasted sequentially from the exposed surface to the design contour, and then the slope boreholes of Fig. 5.2, *a*; and in the second case, all vertical wells are blasted, then one shortened vertical well followed by slope wells on the design contour in sequence Fig. 5.2, *b*; in the third two rows of vertical wells and two rows of inclined wells are blasted in series Fig. 5.2, в. The quality of slope design was assessed by instrumental observations of deformation.

The results of blasting for scarp dressing using this method showed that the scarp top scarp area was traced up to 1 m deep into the massif, which can be easily eliminated by the subsequent dressing of the upper part of the scarp. Absence of sills, presence of traces of contour wells on the slope surface, unchanged granulometric composition in comparison with that of the first scarping scheme allow us to speak about the acceptability of using this method for scarping in the limiting contour.

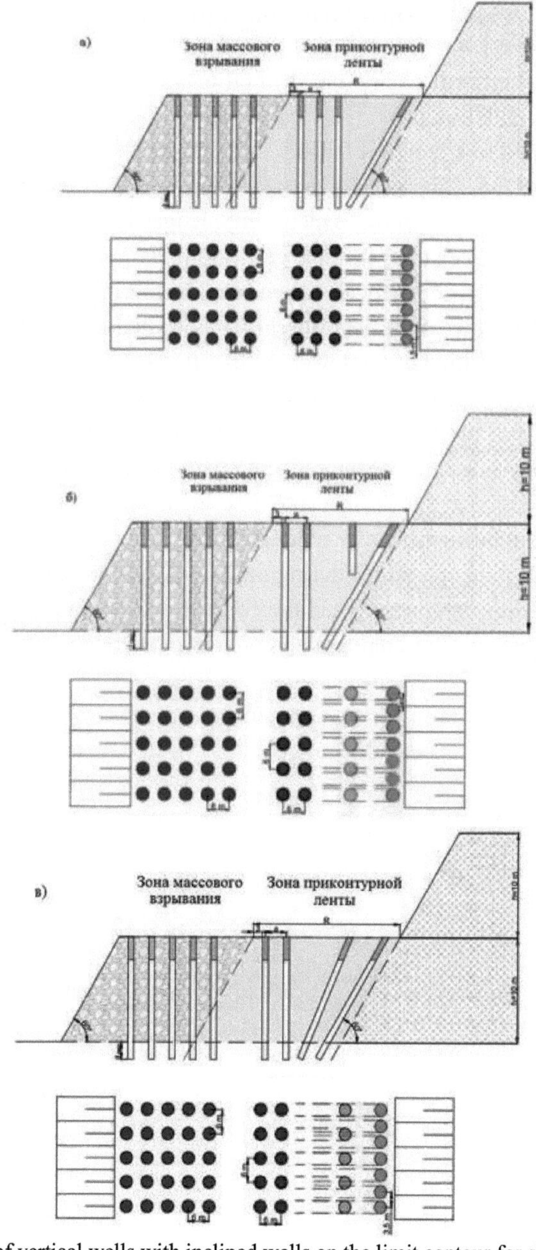

a - three rows of vertical wells with inclined wells on the limit contour for slope design; *b* - two rows of vertical and one shortened row of vertical and one row of inclined wells; *c* - two rows of vertical and inclined wells.

Fig. 5.2. Technological schemes of mining and scarping at the

According to the third scheme the method of scarp development and design by drilling and blasting vertical boreholes, when the last row of short boreholes drilled to a depth of 0.3-0.5 of the scarp height is recommended to be used in areas with large-block rock structure (the average size of the block rib is 1.55 m). The top part of 10 metre high ledges is slanted to a height of 3-5 metres by blasting charges in short vertical boreholes.

Depending on the type of rocks, when placing the upper ledge in the limit contour, the last row of vertical boreholes is drilled at a distance of 3-4 m from the design position of the lower edge of this ledge.

In the work three schemes of installation of the explosive network at multi-row short-delayed blasting at contour belts for maintenance of stability of slopes of quarry sides are considered and recommended (fig. 5.3, 5.4, 5.5).

The above-mentioned shows that dynamic loads on the ledges and sides of the open pit can be significantly reduced by applying a special technology of drilling and blasting operations during mining of the contour protective zone. The extent of the contour zone and the contour protection zone is determined according to established empirical laws in dependence on the strength and fracturing of rocks.

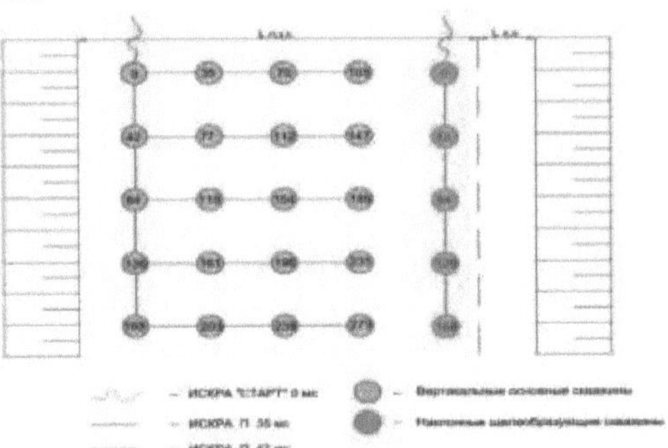

Fig.5.3 Diagonal scheme of multi-row short-delayed explosion

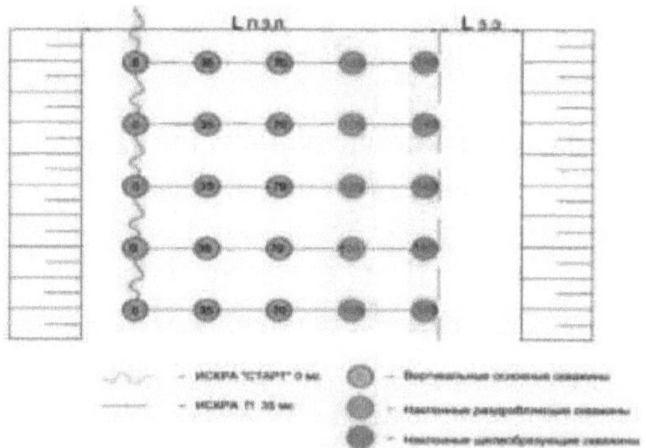

Fig.5.4: Order diagram for multi-row short-delayed explosion

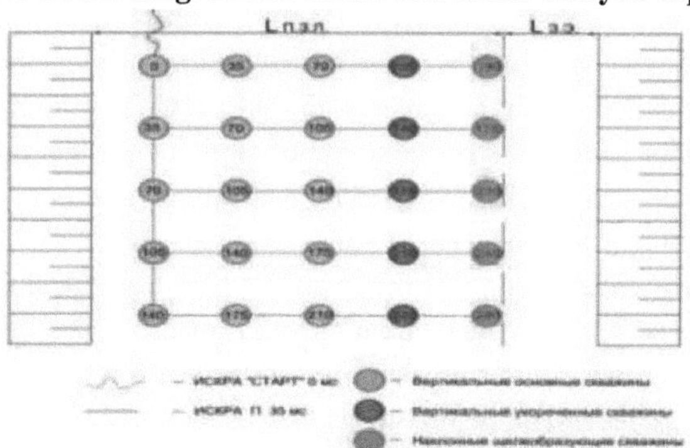

Fig.5.5 Diagonal scheme of multi-row short-delayed explosion

Rock properties are a decisive factor in determining parameters of drilling and blasting operations in the contour zone, selection of technological schemes for mining of the contour protection zone and design of slopes in a stationary position.

Comparative data of experimental explosions are given in Table 5.7. Quantitative assessment of the explosion action indicates the feasibility of the proposed blasting schemes.

Table 5.7

Characterisation of the results of experimental explosions at the quarry Amantaytau

Indicators	Straightforward blasting sequence	Reverse order of blasting
Width of stabbing zone, m	2,8	4
Length of residual deformation zone, m	7	10
Maximum vertical displacement of the array, mm	78	86

Qualitative characterisation of the explosion	Traces of all wells are visible, the slope surface is flat	No traces of contour wells, the slope is uneven

§ 5.3 Development of recommendations to ensure the stability of Amantaytau quarry sides

On the basis of analysis of researches of world practice criterion zoning of instrument massifs of open pit field on stability of slopes, and researches of a modern condition of geomechanical conditions of a rock massif, estimation of the main horizontal stresses on the information on nearby fields-analogues and on geodynamic zoning of deposits.

Also conducted laboratory studies to determine the engineering-geological, physical-mechanical characteristics of rocks and theoretical justification of physical-mechanical parameters of the massif by improved methods of Hook and VNIMI method of finite elements and the method of limit equilibrium established limit parameters of the angle of slope of the sides and blasting technology proposed the following recommendations to ensure the stability of the sides of the North and Central Amantaytau quarries.

1. The recommended zoning of the quarry is based on the slope angles of stable slopes and pit sides, this zoning is valid for the design contour of the quarry. If the design contour is changed, the boundaries and recommended parameters of zones should be revised.
2. In accordance with the performed calculations to determine the general angle of the sides of the quarry, the width of the prism of possible collapse on the eastern side reaches a maximum of 98 m and on the western side of 65 m when adopting the recommended in the work parameters of the sides of the quarry so that the dumps do not have a negative impact on the stability of the sides of the quarry, they should be located at a sufficient distance from the sides of the quarry, that is, outside the prism of possible collapse.
3. Recommendations on the parameters of stable pit ledges have been developed, which are calculated on the basis of information on fracturing with the use of kinematic analysis and special calculations by the limit equilibrium method. It is important to note that the ledge slope angles recommended in this paper can only be achieved by developing and applying special BWR technology to put the ledges in the limit position.
4. Taking into account geomechanical conditions of Amantaytau deposit, namely stratified and relative low strength of rocks, theoretical studies and experimentally recommended the use of special technology of explosive mining of near-contour belts, reducing dynamic loads on the sides of the pit to ensure reliable safety of the contour massif.

5. It is recommended on the basis of the established quantitative and qualitative regularities of massif destruction from explosion the method of definition of width of a near-contour protective zone, the size of a charge in a well, specific expense of explosives, parameters of an arrangement of explosive wells at which safety and long-term stability of a board of a quarry are provided.

Conclusions

1. Quarry zoning by slope angles of stable slopes and quarry sides has been performed for the design contour of the quarry. If the design contour is changed, the boundaries and recommended parameters of the zones should be revised.
2. Theoretical and experimental substantiation of application of special technology of explosive mining of near-contour belts, reduction of dynamic loads on the sides of the open pit to ensure reliable safety of the contour massif.
3. On the basis of the established quantitative and qualitative regularities of massif destruction from explosion the methods of definition of width of a near-contour tape, size of a charge in a well, specific expense of BB, parameters of an arrangement of explosive wells at which reliable safety of a board of a quarry is provided are developed.

CONCLUSION

On the basis of the conducted researches on zoning of sides on stability and substantiation of blasting technology at the contour zone of the quarry the following main conclusions, having theoretical and practical significance, are made:

1. The analysis has shown that the final stage of zoning is the construction of a predictive map, which allocates areas of the board in accordance with the adopted criterion of zoning, predictive map of zoning allows to solve problems associated with the stability of quarry slopes.
2. As a criterion of zoning was taken the degree of stability of the quarry sides by the value of the reserve coefficient, the total displacement of the surface of the instrument massifs, where the horizontal component of the shear vector is predominant.
3. Collection, systematisation and digitisation of exploration drilling data, detailed exploration of the deposit showed that there are no geological sections on potential calculation profiles to assess the stability of pit sides along the most deformation-hazardous directions, which requires additional studies of the rock mass.
4. Theoretical and laboratory studies have established the lateral pressure coefficient, Poisson's coefficient, deformation modulus, physics-mechanical properties of rocks necessary for the assessment of the stability of the sides by the finite element method.
5. The methodology of research of physical and mechanical properties of rocks The obtained values of strength properties of rocks σszh, Pr, adhesion and angle of internal friction allow in the conditions of quarry development to justify the stable parameters of the quarry, both the height of the ledge and the side, as well as the angle of their slope.
6. The results of laboratory studies of determination of tensile strength, deformation properties, adhesion and angle of internal friction, Young's modulus of elasticity and Poisson's coefficient of rocks open possibilities of substantiation of limiting parameters of slope angle of sides and ledges of North and Central Amantaytau quarries.
7. The calculations carried out with the use of nonlinear Hooke-Brown strength criterion and geological strength index (GSI) by finite element method, also with the use of Coulomb strength criterion by limit equilibrium methods strength curves, by different methods, showed their similarity.
8. The stability of the quarry sides in the design contour was assessed by two methods. Modelling by finite element method and calculations by limit equilibrium method (method of vector addition of forces). The results of the

stability assessment signalled insufficient stability of the quarry sides in the design contour. The obtained stability reserve coefficients lie in the range of 0.66-1.06.

9. Theoretical and experimental substantiation of application of special technology of explosive mining of near-contour belts, reduction of dynamic loads on the sides of the open pit to ensure reliable safety of the contour massif.

10. Recommendations for improving the stability of the sides to ensure the efficiency and safety of mining operations of the North and Central Amantaytau open pits were developed.

11. Using the results of the dissertation work in relation to the Amantaytau deposit on the basis of the established data on physical and mechanical properties of the rock massif and structural-geological conditions and performed calculations the selection of the general angles of the quarry sides was carried out, which allows ensuring the normative reserve of stability of the sides at n=1.3, reduce the stripping ratio by 12% and ensure the safety of mining operations by weight
the period of operation of the plant.

LIST OF REFERENCES

1. Bastan P.P. On the possibility of using differences of indicators to assess their variability // Izvestiya Vuzov. Mining Journal. 1963. - №7. - C. 74-84.
2. Bukrinsky V.A., Korobchenko Y.V. Geometrisation of mineral deposits. - M.: Nedra, 1977. - 376 c.
3. Vilesov G.I., Didenko I.M., Ivchenko A.N. Methodology of geometrization of deposits. - Moscow: Nedra, 1973. - 176 c.
4. Gudkov V.M. Theoretical bases of ore homogeneity estimation / V.M. Gudkov, A.A. Vasiliev, K.P. Nikolaev // Mine surveying: Collection of scientific works, Moscow: VZPI, 1976. - Issue 99. - C. 15-28.
5. Margolin A.M. Methods and results of mathematical and statistical research in geological exploration - M.: 1972 - P. 85-113.
6. Myagkov V.F. Geometrization and analysis of geological fields of mineral deposits // Izv. of high schools. Geology and Exploration. Journal, 1984. - № 3. - C. 30-39.
7. Rats M.V. Cracking and properties of fractured rocks. -M.: Nedra, 1970. - 164 c.
8. Ryzhov P.A. Geometry of subsoil / P.A. Ryzhov. Moscow: Uglegletekhizdat, 1952. - 604 c.
9. Sobolevskiy P.K. Modern Mining Geometry / P.K. Sobolevskiy // Scientific Works of MGI: Collection of scientific articles. - M., 1969. - C. 18-63.
10. Ushakov I.N. Mining geometry (geometry of subsoil) State scientific and technical publishing house of literature on mining. - Moscow, 1962. - 459 c.
11. Khokhryakov V.S. Open mining of mineral deposits. - M., Nedra, 1991. - 336 c.
12. Chetverikov, L.I. Theoretical bases of modelling of solid mineral bodies / L.I. Chetverikov. Voronezh: Voronezh University, 1968. - 152 c.
13. Afanasyev B.G. Development of scientific bases for calculation of stability of layered instrument massifs at coal mines : abstract of Cand. Sci. (in Russian): Saint-Petersburg, 1992. - 30 c.
14. Galperin A.M. Geomechanics of open-pit mining. - MOSCOW: MGGU. 2003. - 473 c.
15. Demin A.M., Sokolovskiy M.M. Open mining operations. - Moscow: Uglegletekhizdat, 1958. - 108 c.
16. Development of regulations on stable parameters of ledges and convex sides of the Central and Western quarries and technological schemes of their adjustment. Report / UGI. Ruk. Zoteev V.G. - Ekaterinburg, 1992. - 39 c.
17. Pevzner M.E. Deformations of rocks at quarries / M.E. Pevzner. Pevzner.

- Moscow: Nedra, 1992. - 250 c.
18. Popov V.N. Management of stability of quarry slopes. - Moscow: Gornaya kniga, 2008. - 683 c.
19. Shpakov, P.S.; Popov, I.I. Calculation of pit slopes parameters on the basis of numerical and analytical methods [Text] / P.S. Shpakov, I.I. Popov // Mining Journal, 1988. - №1. - C. 26-28.
20. Sapozhnikov, V.G. K question about the limit height of spoil heaps / V.G. Sapozhnikov // FTPRPI. - 1971. - № 6. - C. 80-86.
21. Turintsev, Yu.I. Methodical guide for determining the maximum angles of redemption of the sides of copper ore pits [Text] / Yu.I. Turintsev, P.V. Koltsov, A.B. Zhabko. - Ekaterinburg: UGGU Publishing House, 2010. - 106 c.
22. Fisenko G.L. Methodical guidelines for determining the angles of inclination of the sides, slopes of ledges and dumps of quarries under construction and in operation / G.L. Fisenko et al. - L.: VNIMI, 1972. - 165 c.
23. Shpakov, P.S. Mine surveying substantiation of geomechanical models and development of numerically analytical methods of calculation of stable quarry slopes [Text]: autoref. doctor of technical sciences. / P.S. Shpakov. - Leningrad, 1988. - 41c.
24. Baron L.I. Lumpiness and methods of its measurement / L.I. Baron. - Moscow: Izd. of the USSR Academy of Sciences, 1960. - 124 c.
25. Belyakov Yu.I. Designing of excavator works / Yu.I. Belyakov. - L.: Nedra, 1983. - 349 c.
26. Golubintsev O. N. Mechanical and abrasive properties of rocks and their drillability / O. N. Golubintsev. - Moscow: Izd-vo "Nedra", 1968. - 198 c.
27. Kutuzov B.N. Destruction of rocks by explosion. Explosive technologies in industry. - M.: MGGU, 1994. - 448 c.
28. Mosinets V.N., Abramov A.B. Fracture of fractured and disturbed rocks. - Moscow: Nedra, 1982. - 248 c.
29. Rakishev B.R. Prediction of technological parameters of blasted rocks at quarries. - Alma-Ata: Nauka, 1983. - 240 c.
30. Rubtsov V.K. Study of structural features of rock massif as applied to blasting // Blasting, 1963. - №53/10. -C. 31-36.
31. Rzhevskiy V.V. Fundamentals of rock physics / V.V. Rzhevskiy, G.Y. Novik. Moscow: Nedra, 1973. - 286 c.
32. Tangaev I.A. Drillability and explosiveness of rocks. - Moscow: Nedra, 1978. - 184 c.
33. Rybin V.V., Potapov D.A., Kalyuzhny A.S. Depth zoning of the open pit field of the Oleniy Ruchey deposit using the geomechanical classification of

Professor D. Lobshire. // Federal State Budgetary Institution of Science Institute of Mining Engineering of the Ural Branch of the Russian Academy of Sciences. Problems of subsoil use, 2014. - 1. - C. 4452.
34. Technological and geomechanical studies of the open pit boundaries of the Oleniy Ruchey deposit: research report (final) under contract No. 2256 dated 30.03.2007 between the Mining Institute of KSC RAS and CJSC "NWPC", funds of the Mining Institute of KSC RAS, Inv. No. 1046 / responsible executor Kozyrev A.A.; executor Reshetnyak S.P., Bilin A.L., Rybin V.V., Lyubin A.N., Churkin O.E., Nagovitsyn O.V., Smagin A.V., Rodina A.V., Kasparian E.V., Zhirov D.V. (responsible for the section). - Apatity: 2007. - 127 c.
35. Report on the results of detailed exploration of the Oleniy Ruchey apatite-nepheline ore deposit in 1980-85 with calculation of reserves as of 1 October 1985 and prospecting and evaluation work on the southwest flank in 1982-1984. RSFSR, Murmansk Oblast. T. 1. Book 1: Text of the report / responsible executor Fanygin A.S. et al. - Apatity: Sevzapzapgeologia, 1985. -354 c.
36. Mountain pressure control in tectonically stressed massifs / Kozyrev A.A. et al. - Apatity, KSC RAS, 1996. - 159 c.
37. Investigation of physical and mechanical properties of the quarry massif and prediction of the stress state of deep horizons based on geological exploration drilling materials to ensure geodynamic safety of the Oleniy Ruchey deposit: research report (2nd stage) under contract No. 2652 between the Mining Institute of KSCRAN and CJSC "NWPC". Kozyrev A.A.; executor. Panin V.I., Rybin V.V., Gubinsky N.O., Potokin M.I., Dannikov I.V., Zhirov D.V., Klimov S.A., Telezhkina N.S., Troshkova A.V. - Apatity, 2010. - 64 c.
38. Laubscher D.H.. A geomechanics classification system for rating of rock mass inmine design / D. H. Laubscher // Journal South African Inst. of Mining and Metallurgy. H. Laubscher // Journal South African Inst. of Mining and Metallurgy. - 1990. - No. 10. - pp. 257-273.
39. Jacubec J., Laubscher D.H.. The MRMR rock mass rating classification system inmining practice. Brisbane. - 2000. - pp. 413-421.
40. Laubscher D.H., Jacubec J. The MRMR Rock Mass Classification for jointed rockmasses. Foundations for Design. Brisbane. - 2000. - pp. 475-481.
41. Nizametdinov F.K., Urdubaev R.A., Ananin A.I., Ozhigina S.B. Methodology for assessing the state and zoning of the sides of deep quarries on the stability factor // Proceedings of the University. "Geotechnologies. Life Safety", 2010. - 4. - c. 44-47.

42. Fisenko G.L. Stability of quarry sides and dumps. - Moscow: Nedra, 1965. - 378 c.
43. Methodical guidelines for determining the angles of inclination of sides, slopes of ledges and dumps of quarries under construction and in operation. - L.: VNIMI, 1972. - 165 c.
44. Temporary methodical instructions on management of stability of non-ferrous metallurgy quarry sides. - M.: MCM SSSR, 1989. - 128 c.
45. Popov I.I., Shpakov P.S., Poklad G.G. Stability of rock dumps. - Alma-Ata: Nauka, 1987. - 225 c.
46. Dunaev V.A., Seriy S.S., Gerasimov A.V., Absatarov S.H. Methodology of zoning of open pit fields on blockiness and explosiveness of rocks of Precambrian basement // Mining information and analytical bulletin. - № 9. 2006. - c. 77-85.
47. Dunayev V.A., Gray S.S. Cracking of metamorphites of the Kursk series in the KMA basin. // "Izvestiya Vuzov. Geology and Exploration", 2003. - №12. - C. 16-18.
48. Tangaev O.P. Drillability and explosiveness of rocks. - Moscow: Nedra, 1978. - 140 c.
49. Temporary classification of rocks by the degree of fracturing in the massif. Interdepartmental Commission on Explosive Engineering // Inf. issue B-199.- M.:IGD, 1968. - C. 31-33.
50. Seriy S.S., Gerasimov A.V. Information technology of geological and mine surveying support and design of drilling and blasting operations at open pits // In: Issues of drainage and protection from waterlogging, mining geology and special mining operations (materials of the 8th International Symposium, part 1). - Belgorod: VIOGEM, 2005. - C.82-86.
51. Koltsov P. V., Ivanov Yu. S., Palutina E. N., Andreeva O. N. Estimation of stability of the sides of the Uchkalinskiy quarry at recultivation. Globus (Geology and Business). - Krasnoyarsk. - №2(46). 2017. - C. 98-102.
52. Rylnikova M.V., Alekseev A.B., Esina E.N. et al. // Mining information-analytical bulletin. № 14. 2011. - C. 79-83.
53. Rules for ensuring slope stability on cuts // SPb: VNIMI. 1998. - 208 c.
54. Hoek E., Carranza-Torres C., Corkum B. Hoek-Brown failure criterion // Toronto: Proc. NARMS-TAC Conference, 200, 1. 2002. - pp. 267-273.
55. Hoek E., Brown E.T. The Hoek-Brown failure criterion and GSI // Journal of rock mechanics and geotechnical engineering. 2018. - P. 306.
56. Richard E. Goodman, Genhua Shi. Block Theory and Its Application to Rock Engineering // 1985. - P. 352.
57. Guidelines for the design of quarry faces [Text] / [Peter Stacey et al.];

eds: John Reed, Peter Stacey ; [translated from English: A. S. Benthen] // Ekaterinburg: Pravoved: Polymetal. 2015. - 527 c.
58. Gordeev V.A., Samarin A.V. Criterion for zoning the quarry field by slope stability // Proceedings of the Ural State Mining University. Issue 3, 1993. - C. 70-71.
59. Gordeev, V.A.; Samarin, A.V. Score system for estimating the stability of quarry slopes // Problems of improving the efficiency of surveying works at mining enterprises: Interuniversity scientific collection / edited by V.A. V.A.; V.A. Kuznetsov. : B. A. Gordeev (editor-in-chief) et al. - Ekaterinburg: UGI, 1992. - 68 c.
60. Summary reserves estimation for the Amantaitau gold deposit. Detailed exploration of the deposit. Samarkandgeologiya State State Enterprise. Daugyztau GRE, Amantaytau GRP. - 1994. - 90 c.
61. Silkin A.A. et al. "Management of long-term stability of slopes at quarries in Uzbekistan". Tashkent, ed. "Fan" -2005. - 160 c.
62. Methodical guidelines for determining the angles of inclination of sides, slopes of ledges and spoil dumps of quarries under construction and in operation. - L.: VNIMI, 1972. - 12 c.
63. Rules for ensuring slope stability at coal mines. - Izd. VNIMI, St. Petersburg, 1998. - 136 c.
64. Aitmatov. I.T. "Geomechanics of ore deposits of Central Asia". - Frunze, Academy of Sciences of the Kyrgyz SSR, 1987. - 247c.
65. Temporary methodical instructions on management of stability of non-ferrous metallurgy quarry sides. - Moscow: "Unipromed", 1989. - 22 c.
66. Fisenko G.L. "Stability of quarry sides and dumps". - Moscow: "Nedra", 1965. - 203 c.
67. GOST 12071-2014. Soils. Selection, packing, transport and storage of samples. - M.: 2015. - 32 c.
68. GOST 20522-2012. Soils. Method of statistical processing of test results. - M.: 2013. - 28 c.
69. GOST 25100-2011. Classification. - M.:, MNTX, 2013. - 16 c.
70. GOST 5180-2015. Soils. Methods of laboratory determination of physical characteristics. - M.: 2016. - 18 c.
71. GOST 21153.2-84. Rocks. Methods of determination of uniaxial compression strength. - M.: 1984. - 24 c.
72. GOST 21153.3-85. Rocks. Methods of determination of uniaxial tensile strength. - M.: 1985. - 36 c.
73. GOST 24941-81 Rocks. Methods of determination of mechanical properties by loading with spherical indenters. - M.: 1981. -28 c.

74. GOST 12248-96. Soils. Methods of laboratory determination of strength and deformability characteristics. - M.: 2011. - 36 c.
75. GOST 21153.8-88. Rocks. Methods of determination of bulk compressive strength. - M.: 1988. 17 c.
76. GOST 8269.0-97 Crushed stone and gravel from dense rocks and industrial wastes for construction works.Methods of physical and mechanical tests. - M.:1997. - 18 c.
77. GOST 21153.5-88 Rocks. Method of determination of compressive shear strength. - M.: 1988. - 22 c.
78. GOST 21153.7-75 Rocks. Method of determination of velocities of elastic longitudinal and transverse waves propagation. - M.: 1976. - 24 c.
79. RSN 51-84. Engineering surveys for construction. Production of laboratory studies of physical and mechanical properties of soils. - M.: 1984. - 122 c.
80. Litvinov field laboratory PLL-9. Filtration coefficient of clayey soils. - St. Petersburg, 2012. - 16 c.
81. Properties of rocks and methods of their determination. - Edited by M.M. Protodyakonov. - M.: 1984. - C. 116-118.
82. Methodical guidelines for determining the angles of inclination of sides, slopes of ledges and spoil dumps of quarries under construction and in operation. - L.: VNIMI, 1972. - 36 c.
83. Research and Development Report "Investigation of physical and mechanical properties of rocks of Northern and Central Amantaytau deposits and justification of limiting parameters of the slope angle of the sides and ledges of open pits". 1 stage. JSC VNIMI, St. Petersburg, 2019. - 64 c.
84. Research and Development Report "Study of physical and mechanical properties of rocks of the Northern and Central Amantaytau deposits and justification of the limiting parameters of the slope angle of the sides and ledges of open pits". 2 stage. JSC VNIMI, St. Petersburg, 2019. - 42 c.
85. E. Hoek, E.T. Brown. The Hoek-Brown failure criterion and GSI - 2018 edition // Journal of Rock Mechanics and Geotechnical Engineering. 2018. - pp. 119.
86. E. Hoek. Putting numbers to geology - an engineer's viewpoint // Quarterly Journal of Engineering Geology and Hydrogeology, 32. 1999. - pp. 119.
87. Hoek, E and Karzulovic, A. Rock mass properties for surface mines, in Slope Stability in Surface Mining, (Edited by W.A. Hustralid, M.K. McCarter and D.J.A. van Zyl), Littleton, Colorado: Society for Mining, Metallurgical and Exploration (SME), 2000. - pp. 59-70.

88. Guidelines for open pit slope design. Guidelines for open pit slope design: Scientific edition // edited by D. Reed, P. Stacey; English translation by A. S. Bentshen; scientific editor A. B. Makarov. Ekaterinburg: Pravoved, 2015. -528 c.
89. Popov V.N., Baikov B.N. Technology of quarry sidetracking. - Moscow: Nedra, 1991. - 252 c.
90. Rakishev, B.R. Rational parameters of charge arrangement in a ledge // Blasting, 2009. - C. 81-90.

Attachment

Annex 1

Technological parameters for drilling additional wells

№ borehole.	Profile	Design depth, m	Projected (expected) geological section	Test interval	Min. drilling diameter, mm
T-1	3	150	0-60 - Clay hard light grey, yellow, with thin marl interlayers; 60-75 - Light grey sand (sandstone) dense low-moisture; 75-140 - Clay is hard; mudstone with sandstone interlayers; 140-150 - Quartzed sandstone, durable	Determine the depth of the roof of Palaeozoic rocks, without sampling	89
T-2	5	130	0-10 - Clay hard light grey; 10-20 - Mergel; 20-40 - Clay is hard; 40-50 - Mergel; 50-80 - Sand of unspecified density; 80-125 - Hard argillite-like clay; 125-130 - Quartzed sandstone	Monoliths: 10,25,30,40. (clays) 15,45 (marl) 55,60,65,70,75,80 (sands)130 (quartz sandstone)	112-89 (at the face)
T-3	6	140	0-55 - Clay is hard; 55-85 - Sand is loose; 85-135 - Hard clay with sandstones; 135-140 - Sandstone quartzed rugged sandstone	Establish the roof of Palaeozoic rocks, without sampling	89
T-4		90	0-50 - Clay is hard; 50-85 - Sand is loose; 85-90 - Hard clay with thin marl interlayers	Monoliths: 20,40,50 (clays) 60,65,70,75,80,85 (sands)	112-89 (at the face)
T-5	7	120	0-80 - Clay is hard; 80-95 - Mergel; 95-120 - Fine-grained sandstone Mz-Kz	Monoliths: 20,40,60,80 (clays) 85,90 (marl) 100,120 (sandstone)	112-89 (at the face)
T-6	9	90	0-80 - Clay is hard; 80-90 - Fine-grained sandstone	20,40,60,80 (clays) 90 (sandstone)	112-89 (at the face)
T-7	10	50	0-45 - Hard clay; 45-50 - Siltstone	Establish the roof of Paleozoic rocks, without sampling	89

Annex 2

List of wells required to create a 3D model

fields

Table 1- Wells of the North field

Well numbers:							
1300	1403	1644	1703	1800	1902	2006	139
1301	1404	1647	1704	1802	1908	2009	283
1305	1416	1648	1705	1804	1915	2013	293
1306	1421	1649	1706	1805	1916	2123	544
1310	1424	1655	1707	1812	1917	2126	739
1311	1425	1658	1715	1819	1932	2128	
1314	1426	1661	1716	1822	1938	2130	
1318	1427	1664	1722	1824	1942	2132	
1324	1428	1666	1723	1826	1944	2164	
1327	1429	1667	1725	1830	1951	2175	
1328	1437	1671	1729	1840	1952	2305	
1330	1447	1673	1734	1841	1990a		
1331	1451	1674	1753	1845	1998		
1332	1456	1676	1755	1854	1999		
1343	1460	1678	1763	1856			
1348	1462	1679	1764	1861			
1350	1465	1680	1765	1862			
1351	1467	1682	1770	1867			
1358	1468	1684	1771	1877			
1361	1472	1686	1783	1879			
1377	1473	1688	1784	1883			
1379	1476	1690	1786	1885			
1381	1487	1691	1787	1890			
1381	1490	1692	1788				
1382	1491	1697	1791				
1391	1498	1698	1791				
1393	1499	1698	1799				
1398							
1399							
Total: 158 wells							

Table 2- Wells of the Central field (North-East, East, South)

№ profiles	№№ wells										Number of wells.
prof.47	183	169	662	664	689						5
prof.49	681	1574	1575								3
prof.50	682	1573	2112								3
prof.51	338	1339									2
prof.52	178	1335	1569	2113							4
Prof. 53	185										1
Prof. 54	186	586	235	1556							4
prof.55	187	320	510								3
Prof. 56	300	332	342	1564							4

Prof. 57	189	301	1562	2109						4
Prof. 58	190	302	1560							3
Prof. 59	191	364	365							3
Prof. 82	790	791								2
Prof. 83	788	789	2263							3
Prof. 84	1526	1550	1965	2254	2255	2257				6
prof.85	1527	2247								2
Prof. 86	333	335	1525	2237	2240					5
Prof. 87	1519	2229	2230							3
prof.88	328	329	887	1518	2223	2224	2228			7
Prof. 99	193	194	195	316	322	1514	1516	1986		8
prof. 90	330	1422	1913	2211	2213					5
prof.91	240	245	255	982	1520	1554	1559	1614	1960	9
prof. 92	132	227	306	980	984	1586	1790	8005	8006	9
prof. 92	974	1505	1553	1557	1941	1969	1972	2105	8003	8004	10
prof. 94	969	970	994	995	1506	1507	1558	1574	1839	1945	12
	8001	8002									
prof.95	148	154	198	278	360	1895	1966	1994			8
prof.96	155	803	967	996	997	1366	1390	1581	1946	1948	11
	1994										
prof.97	150к	230	317	318	387	717	1905	1920	2027	2028	11
	2029										
prof.98	158	360	491	546	665	1317	1548	1621	1623	2145	10
prof.99	159	372	373	492	714	715	716	1544	1545	2141	10
prof.100	100	464	493	546	1541	1542	1543	1570	2139		9
prof.101	128	337	538	539	582	648	694	800	801	1539	12
	1606	2138									
prof.102	540	553	578	579	580	581	615	802	1360a	1537	12
	1921	2137									
1	2										3
prof.103	369	439	805	1536a	1995						5
prof.104	260	277	287	437	599	637	1594	1615	1618	1811	10
prof.105	436	590	643	690	805	807	1534				7
Prof. 106	288	470	495	531	600	603	937	938	1588	1909	11
	1922										
Prof. 107	498	534	545	636	708	892	939	940	942	1291	14
	1532	1595	1601	2117							
prof.108	530	759	815	944	945	1450	1594	1928	2118	2120	10
Prof. 109	249	482	587	588	589	644	730	731	732	741	13
	1530	1609	1625								
prof.110	261	294	497	535	563	564	565	572	592	1923	10
prof.111	248	595	596	645	652	653	654	655	656		9
Prof. 112	250 533	257 556	264 610	295 611	485	487	512	517	519	528	14
prof.113	608	884									2
Prof. 114	570	569	568	567	613	612	526	551	552	525	13
	885	561	253								
prof.115	2110										1

prof.116	511	524	583								3
Total number of wells.											325

Table 3- Wells of the South-Western part of the Central Section

profile no.	No. of wells					Number of wells.
prof. 92	673	674				2
prof. 94	331	671	672			3
prof.95	327					1
prof.96	422	686				2
prof.97	668	669	670	689		4
prof.98	170	228	452	453	454	5
prof.99	635	663	678			3
prof.101	314	677				2
Total number of wells.						22

Annex 3

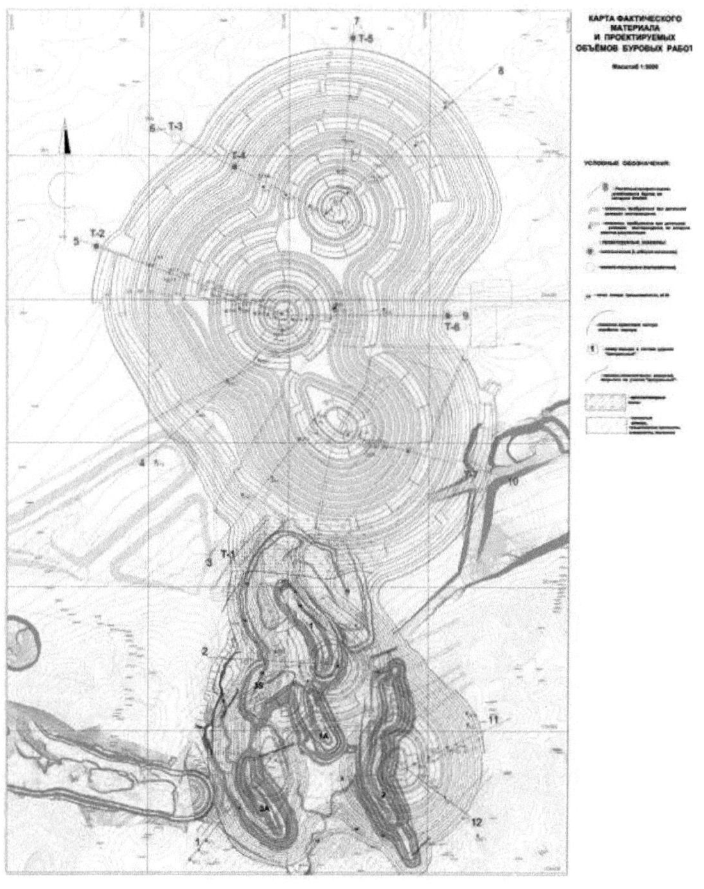
КАРТА ФАКТИЧЕСКОГО
МАТЕРИАЛА
И ПРОЕКТИРУЕМЫХ
ОБЪЁМОВ БУРОВЫХ РАБОТ
7
T-5
8
6
T-3
T-4
T-2
5
9
T-6
4
10
3
T-1
2
11
12
1

Annex 4

Midas GTS NX Certificate of Conformity

СИСТЕМА СЕРТИФИКАЦИИ ГОСТ Р

ФЕДЕРАЛЬНОЕ АГЕНТСТВО ПО ТЕХНИЧЕСКОМУ РЕГУЛИРОВАНИЮ И МЕТРОЛОГИИ

СЕРТИФИКАТ СООТВЕТСТВИЯ

№ РОСС KR.HB61.H05884

Срок действия с 30.04.2020 по 29.04.2023

№ 0467208

ОРГАН ПО СЕРТИФИКАЦИИ RA.RU.11HB61
Орган по сертификации ООО "ЦЕТРИМ". Адрес: 153000, РОССИЯ, Ивановская область, город Иваново, улица Богдана Хмельницкого, дом 36В. Телефон +7 4932773165. Адрес электронной почты info@cetrim.ru

ПРОДУКЦИЯ Программные комплексы для расчета и проектирования конструкций различного назначения и выполнения комплексных геотехнических расчетов midas GTS NX / FEA NX (в трехмерной и плоской постановках) и midas SoilWorks (в плоской постановке) согласно Приложению, бланки №0095509-095512. Серийный выпуск.

код ОК
58.29.29.000

СООТВЕТСТВУЕТ ТРЕБОВАНИЯМ НОРМАТИВНЫХ ДОКУМЕНТОВ
ГОСТ 28195-89, разд. 2, п.2.1 (пп. 1.1, 1.2, 2.1, 2.2, 2.3, 3.1, 3.2, 3.3, 6.1, 6.2); ГОСТ 28806-90, разд. 2, пп. 13-16; ГОСТ Р ИСО/МЭК 9126-93, разд. 4, пп. 4.1-4.4; ГОСТ Р ИСО 9127-94, разд. 6, пп. 6.1, 6.3-6.5; ГОСТ Р ИСО/МЭК 12119-2000, разд. 3, пп. 3.1.1, 3.1.3,, 3.1.4, 3.1.5, 3.2.1-3.2.5, 3.3.1, 3.3.2, 3.3.3; ГОСТ 27751-2014; ГОСТ 25100-2011; ГОСТ 5180-2015; ГОСТ 12248-2010; ГОСТ 20276-2012; нормативных и программных документов см. Приложение, бланки №0095509-095512

код ТН ВЭД
-

ИЗГОТОВИТЕЛЬ MIDAS Information Technology Co., Ltd. Адрес: 463-400, КОРЕЯ, РЕСПУБЛИКА, MIDAS IT Tower – Pangyo Seven Venture Valley, 633 Sampyeong-dong Bundang-gu, Seongnam-si, Gyeonggi-do, телефон: +82-31-789-1955, адрес электронной почты: info@midasit.com.

СЕРТИФИКАТ ВЫДАН Общество с ограниченной ответственностью "МИДАС", ОГРН: 1137746856565, ИНН: 7736664814, КПП: 772501001. Адрес: 115280, РОССИЯ, г. Москва, Ул. Ленинская Слобода, д. 19, комната 21К, телефон: +7 (495) 269-0257, адрес электронной почты: rusupport@midasit.com.

НА ОСНОВАНИИ
Протокол испытаний № 003/Z-30/04/20 от 30.04.2020 года, выданный Испытательной лабораторией Общества с ограниченной ответственностью "ТАНТАЛ" (аттестат аккредитации РОСС RU.31578.04ОЛН0.ИЛ13)

ДОПОЛНИТЕЛЬНАЯ ИНФОРМАЦИЯ
Схема сертификации: 3с

М.П.

Руководитель органа — П.Г. Рухлядев

Эксперт — В.П Широков

Сертификат не применяется при обязательной сертификации

Appendix 5.
Results of numerical modelling

a)

б)

FEM modelling results for section 1-1: a) highest shear strains; b) highest tangential stresses

б)

FEM modelling results for section 2-2: a) highest shear strains; b) highest tangential stresses

a)

б)

FEM modelling results for section 3-3: a) highest shear strains; b) highest tangential stresses

a)

б)

FEM modelling results for section 4-4: a) largest shear strains; b) largest tangential stresses

a)

б)

FEM modelling results for section 5-5: (a) the largest shear deformations; b) highest tangential stresses

a)

б)

FEM modelling results for section 6-6: a) highest shear strains; b) highest tangential stresses

a)

б)

FEM modelling results for section 7-7: a) highest shear strains; b) highest tangential stresses

а)

б)

FEM modelling results for section 8-8: a) largest shear strains; b) largest tangential stresses

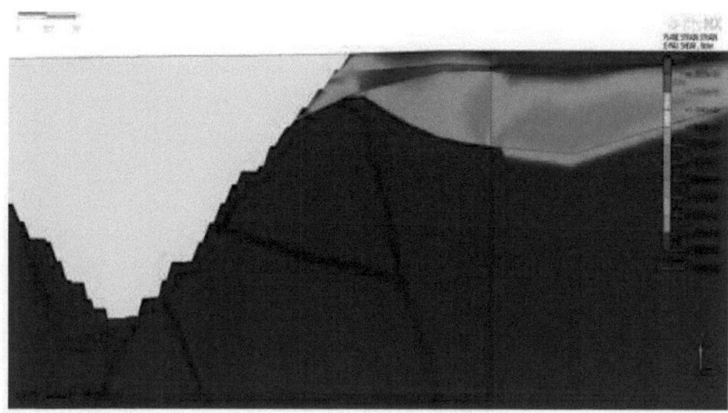

б)

FEM modelling results for section 9-9: a) highest shear strains; b) highest tangential stresses

a)

б)

FEM modelling results for section 10-10: a) highest shear strains; b) highest tangential stresses

a)

б)

FEM modelling results for section 11-11: a) highest shear strains; b) highest tangential stresses

a)

б)

FEM modelling results for section 12-12: a) largest shear strains; b) largest tangential stresses

Annex 6
Results of MVSS stability calculations

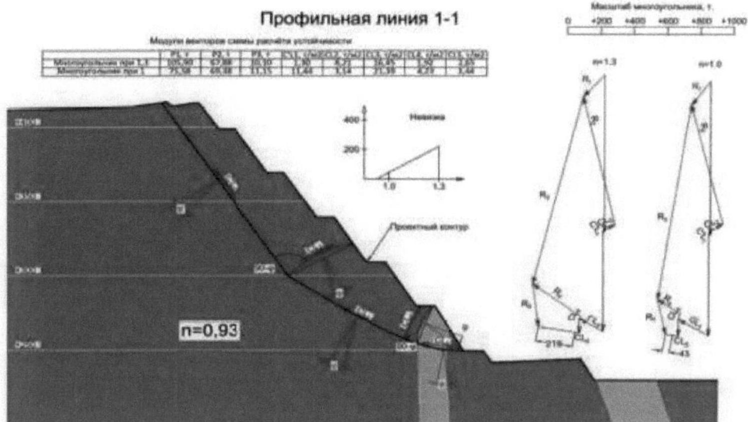

Stability calculation for section 1-1

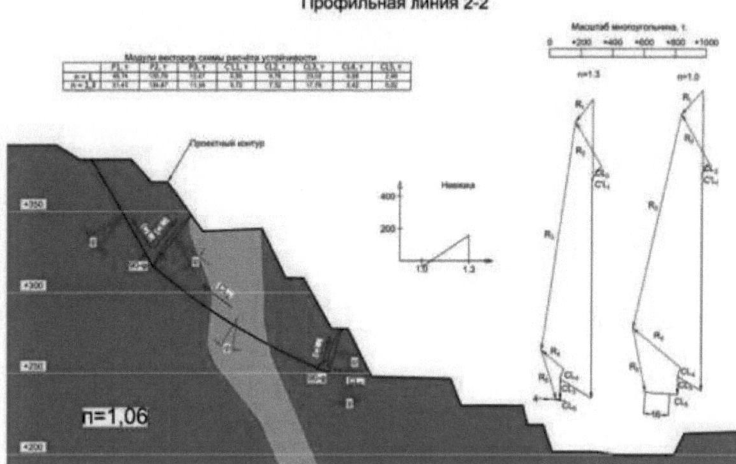

Stability calculation for section 2-2

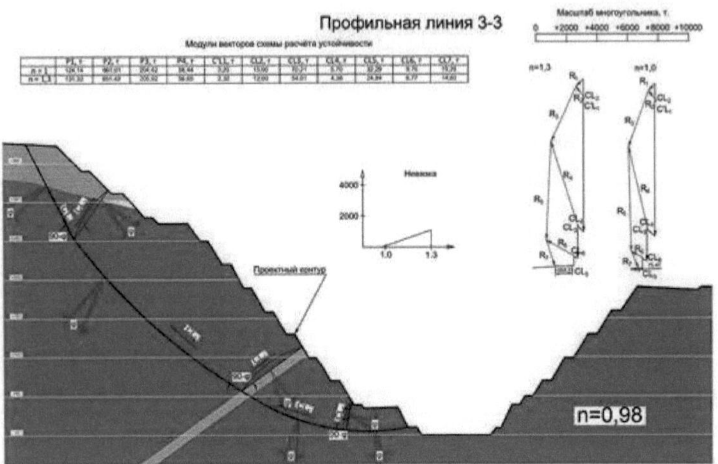

Stability calculation for section 3-3

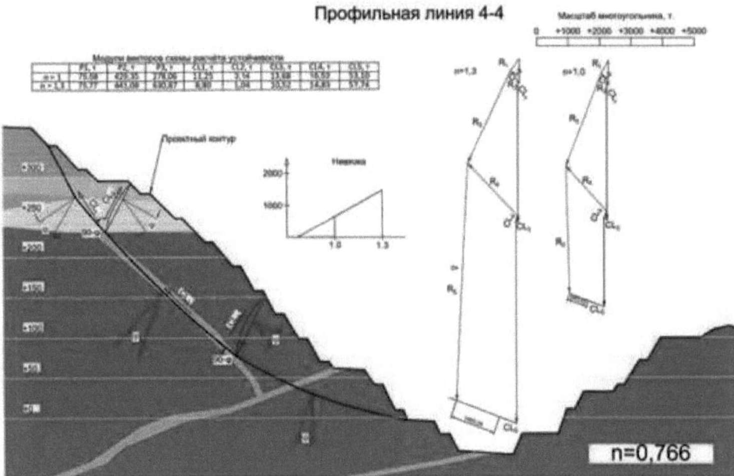

Calculation of stability in section 4-4

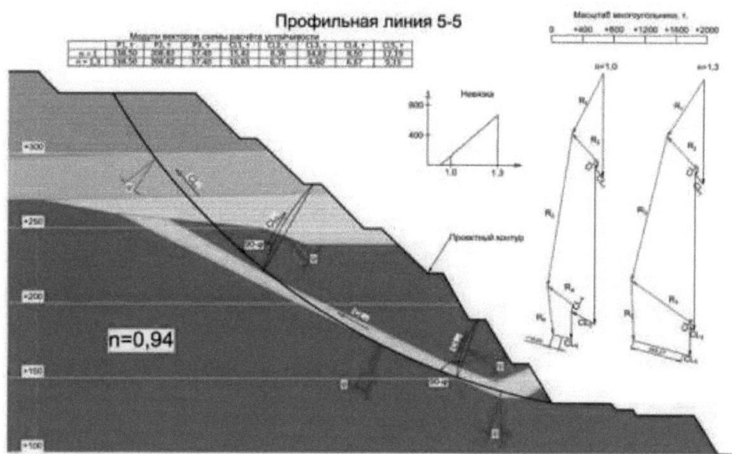

Calculation of stability in section 5-5

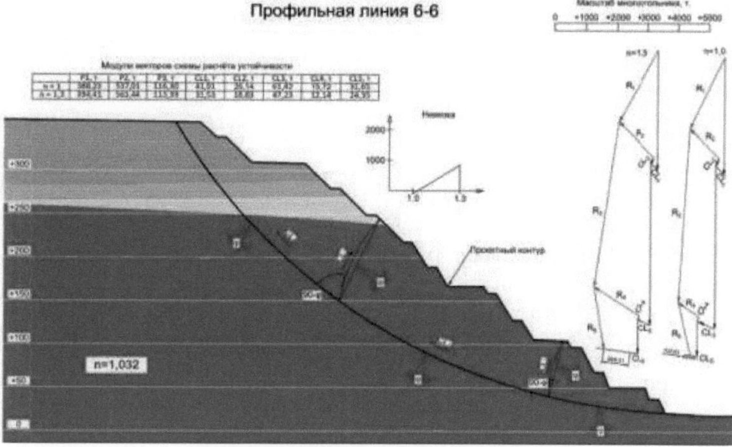

Stability calculation for section 6-6

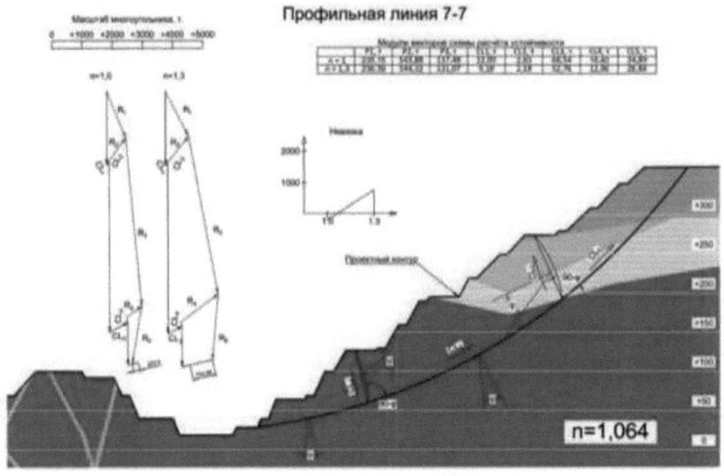

Stability calculation for section 7-7

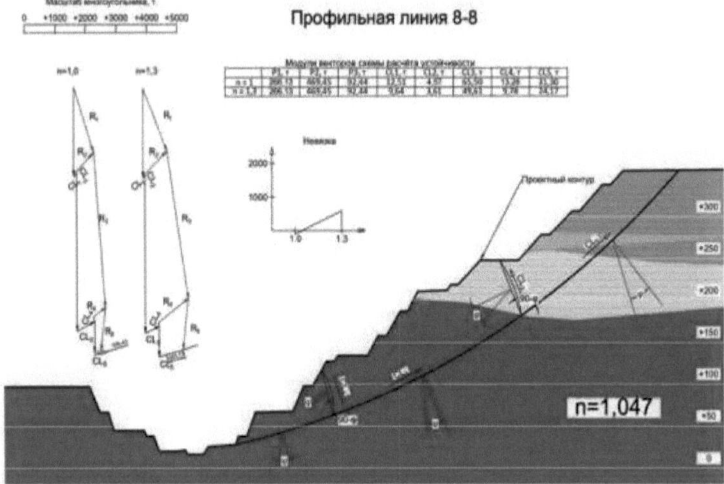

Stability calculation for section 8-8

Профильная линия 9-9

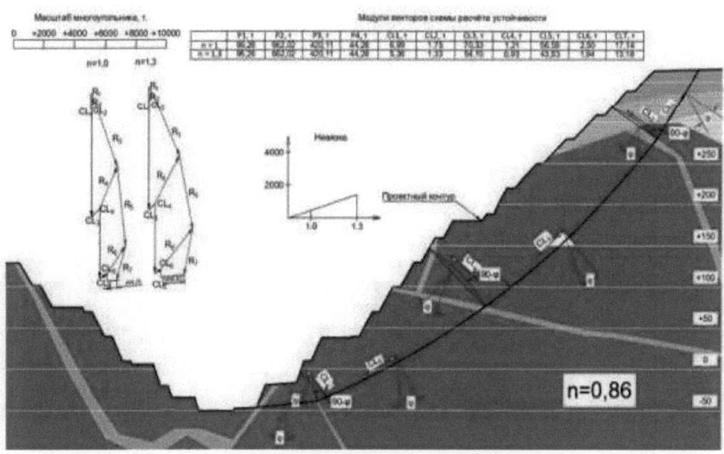

Stability calculation for section 9-9

Профильная линия 10-10

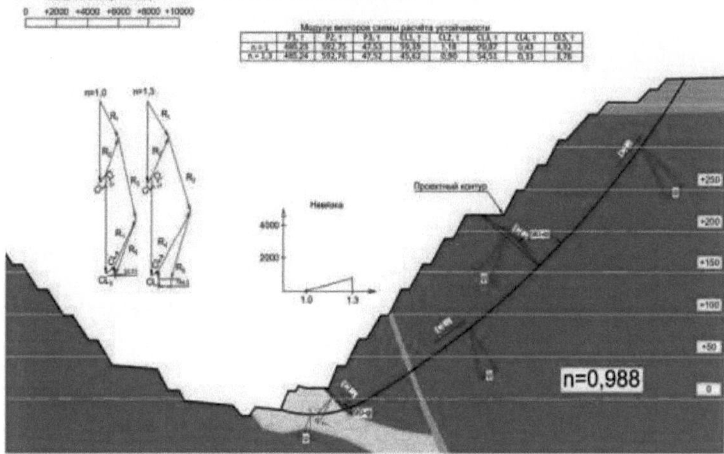

Stability calculation for section 10-10

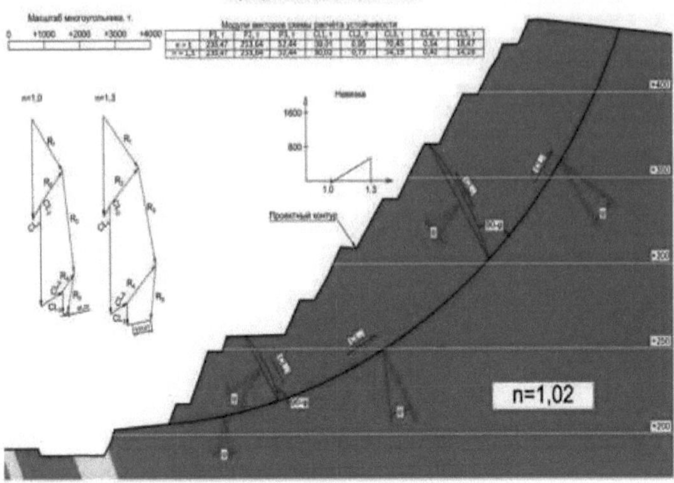

Stability calculation for section 11-11

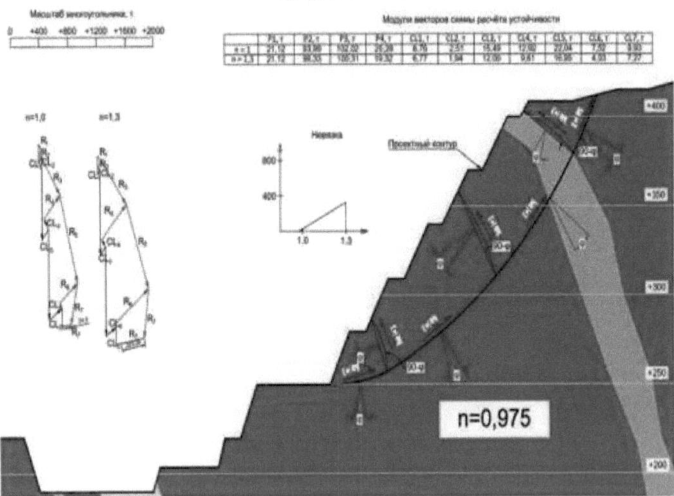

Stability calculation for section 12-12

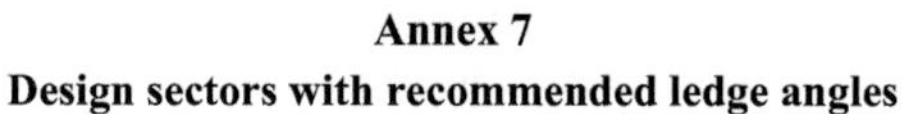

Annex 7
Design sectors with recommended ledge angles

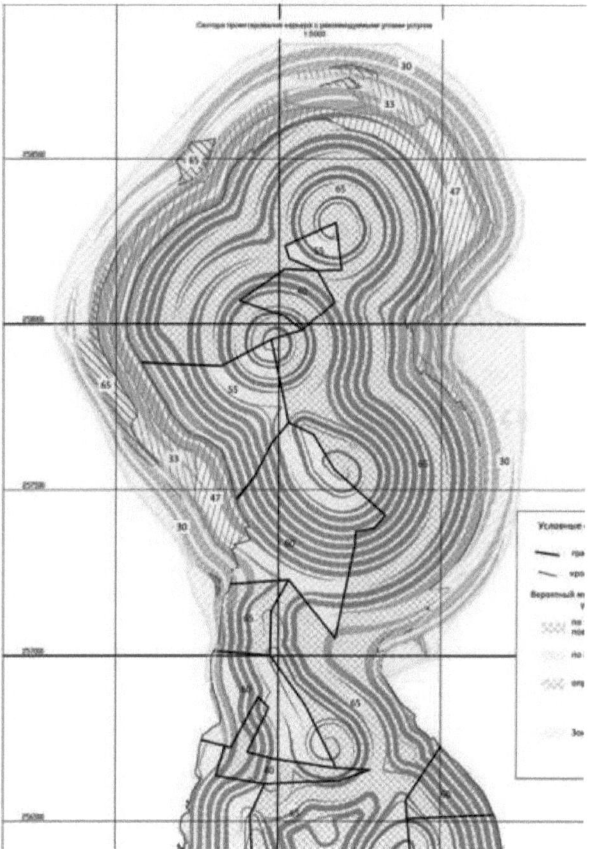

Annex 8
Quarry zoning by general corners

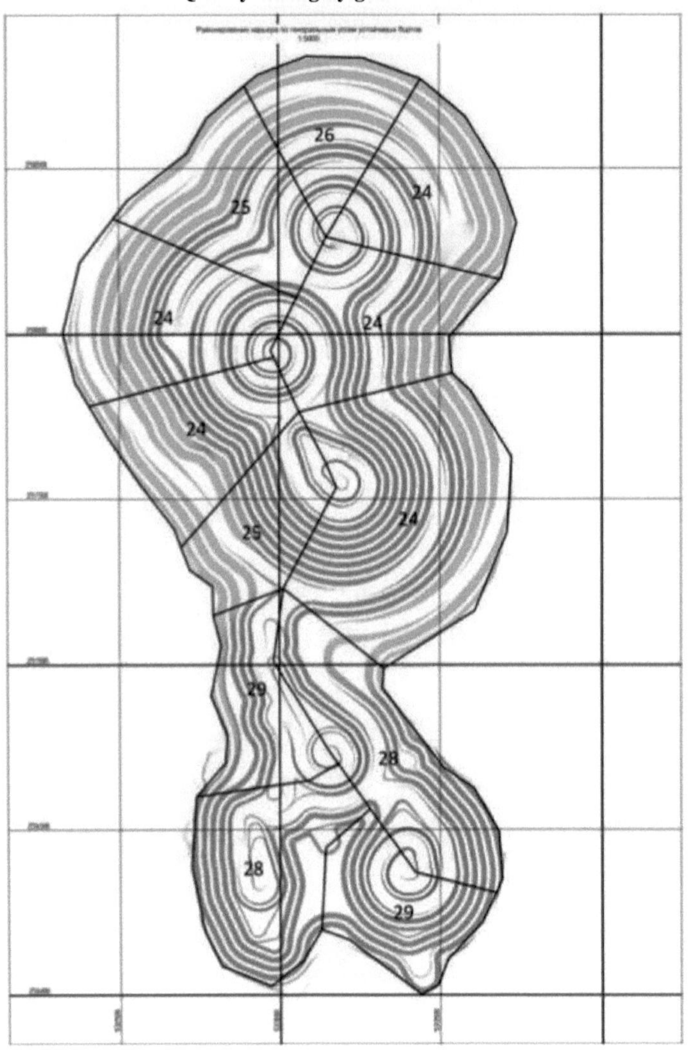

Printed by Books on Demand GmbH, Norderstedt / Germany